Pacific Partnerships for Health

Charting a Course for the 21st Century

Committee on Health Care Services in the
U.S.-Associated Pacific Basin

Division of Health Care Services and
Board on International Health

INSTITUTE OF MEDICINE

Jill C. Feasley and Robert S. Lawrence, *Editors*

NATIONAL ACADEMY PRESS
Washington, D.C. 1998

NATIONAL ACADEMY PRESS • 2101 Constitution Avenue, N.W. • Washington, DC 20418

NOTICE: The project that is the subject of this report was approved by the Governing Board of the National Research Council, whose members are drawn from the councils of the National Academy of Sciences, the National Academy of Engineering, and the Institute of Medicine. The members of the committee responsible for the report were chosen for their special competences and with regard for appropriate balance.

This report has been reviewed by a group other than the authors according to procedures approved by a Report Review Committee consisting of members of the National Academy of Sciences, the National Academy of Engineering, and the Institute of Medicine.

The Institute of Medicine was chartered in 1970 by the National Academy of Sciences to enlist distinguished members of the appropriate professions in the examination of policy matters pertaining to the health of the public. In this, the Institute acts under both the Academy's 1863 congressional charter responsibility to be an adviser to the federal government and its own initiative in identifying issues of medical care, research, and education. Dr. Kenneth I. Shine is president of the Institute of Medicine.

Support for this project was provided by the Health Resources and Services Administration, U.S. Department of Health and Human Services, with additional support provided by the Archstone Foundation and the U.S. Department of the Interior. The opinions expressed in this report are those of the Committee on Health Care Services in the U.S.-Associated Pacific Basin and do not necessarily reflect the views of the funders.

International Standard Book Number: 0-309-05948-8

Additional copies of *Pacific Partnerships for Health: Charting a Course for the 21st Century* are available for sale from the National Academy Press, 2101 Constitution Avenue, N.W., Box 285, Washington, DC 20055. Call (800) 624-6242 or (202) 334-3313 (in the Washington metropolitan area) or visit NAP's on-line bookstore at www.nap.edu.

For more information about the Institute of Medicine, visit the IOM home page at www.nas.edu/iom.

Copyright 1998 by the National Academy of Sciences. All rights reserved.

Printed in the United States of America

The serpent has been a symbol of long life, healing, and knowledge among almost all cultures and religions since the beginning of recorded history. The serpent adopted as a logotype by the Institute of Medicine is a relief carving from ancient Greece, now held by the Staatliche Museen in Berlin.

Cover artwork: *Women and Children*, by Valerie Hunton. Cover design by Francesca Moghari.

COMMITTEE ON HEALTH CARE SERVICES IN THE U.S.-ASSOCIATED PACIFIC BASIN

Robert S. Lawrence* (*Chair*), Associate Dean for Professional Education and Programs and Professor of Health Policy, School of Hygiene and Public Health, Johns Hopkins University, Baltimore, Maryland

Dyanne D. Affonso,* Dean and Professor, Nell Hodgson Woodruff School of Nursing, Emory University, Atlanta, Georgia

Carolyne K. Davis,* International Health Care Consultant, Ernst and Young, Washington, D.C.

William H.J. Haffner, Professor and Chair, Department of Obstetrics and Gynecology, Uniformed Services University of the Health Sciences, Bethesda, Maryland

Glen E. Haydon, International Consultant, Des Moines, Iowa

Francis X. Hezel, Director, Micronesian Seminar, Chuuk, Federated States of Micronesia

Dean T. Jamison,* Professor, Center for Pacific Rim Studies, University of California at Los Angeles, Los Angeles, California (resigned from committee in July 1997)

Agnes Manglona McPhetres, President, Northern Marianas College, Saipan, Commonwealth of Northern Marianas Islands

Hon. Tosiwo Nakayama, Vice President for Governmental Affairs, Bank of Guam, Chuuk, Federated States of Micronesia

Paul W. Nannis, Commissioner of Health, City of Milwaukee Health Department, Milwaukee, Wisconsin

Terence A. Rogers, Dean Emeritus, John A. Burns School of Medicine, University of Hawaii, Honolulu, Hawaii

David N. Sundwall, President, American Clinical Laboratory Association, Washington, D.C.

Study Staff

Jill Feasley, Program Officer
Annice Hirt, Research Assistant
Heather Callahan, Project Assistant
Evelyn Simeon, Administrative Assistant
Clyde Behney, Deputy Executive Officer
Christopher Howson, Director, Board on International Health

---

*Institute of Medicine member.

Preface

The people of the U.S.-Associated Pacific Basin live half a world away from me and from the health policy makers in Washington, D.C. Yet, we are all linked through decades-old economic, legal, social, and cultural ties and a shared aspiration for better health for our families and children.

Members of the committee that the Institute of Medicine convened to carry out the project described in this report are from several parts of the United States, and three members currently live in the region itself. Some of us had relatively little knowledge of the region; others had spent years living and working to improve the overall well-being and health care of people living on the islands of the Pacific Basin. All of us were able to visit at least two of the jurisdictions during our site visits. We were struck by the tremendous differences between the health care services available to most people in the 50 U.S. states compared to those in the Pacific Basin region. Even within the region, the differences were striking: the relative abundance of services and providers in Guam and the Commonwealth of the Northern Mariana Islands compared with the dearth of either in places like Chuuk.

Yet, we also saw similarities. The damage and suffering wrought by alcohol and substance abuse, unhealthy diets, and unintended injuries are the same—whether they occur in Milwaukee, Majuro, Guam, or Georgia. On both sides of the Pacific, people at all levels are trying to figure out how to spend health care dollars more wisely. We all struggle with how to make available and provide access to certain basic health care services for the entire population, because we know that some of our vulnerable citizens are falling through the holes in the safety net. We recognize the need to shift our health care focus away from disease treatment and toward disease prevention and health promotion, but we lack the will to reallocate the necessary funds. In many respects, we are not so very different at all.

This report charts a course for health care services in the region for the coming years. It calls for a strengthening of community-based primary care, better coordination of efforts between the United States and the island jurisdictions, greater involvement of local communities and individuals in promoting health, and improved education and training for the health care workforce.

Change is rarely easy. Some of the recommendations that we have made will require the U.S. government and the island jurisdictions to make very difficult decisions about what is truly important and which activities can no longer be supported. Our report comes at a time, however, when discussions about these decisions have already begun. Some communities are farther along and are better-equipped than others to make these meaningful changes. The committee and I believe that the recommendations contained in this report, if adopted, will make a substantial contribution toward the goal of current health reforms to create healthy islands and island populations for many years to come.

Robert Lawrence, M.D.
Chair

Acknowledgments

The committee and staff are indebted to many individuals who provided a great deal of help during the study: our project officers from the Health Resources and Services Administration, Anne Chang and Tom Coughlin, who were aided by Howard Lerner and LuAnn Pengidore; Mary Ellen Kullman Courtright from the Archstone Foundation; and Darla Knoblock from the U.S. Department of the Interior. A special thank you to Roylinne Wada, who diligently provided critical information, data, and analysis throughout the project.

We also thank the speakers from our first meeting: Connie Arvis, Tom Bell, Joe Iser, and Greg Dever. We would also like to recognize the many participants in our meeting in Saipan: Father Roger Tenorio, Eliuel Pretrick, Jesse Borja, Victor Yano, Joe Flear, Vita Skilling, Anamarie Akapito, Jimione Samisoni, Wame Baravilala, June Shibuya, Isamu Abraham, Susan Schwartz, Marcus Samo, Elena Scragg, and Joe Villagomez. Jim Johnson, Joe Flear, and Patricia Ruez made presentations at the final committee meeting in August 1997.

Before the Saipan meeting in April, the committee and staff visited each jurisdiction and met with literally hundreds of individuals. We have listed many of them below; we regret that we have failed to include each and every person who helped make our stays so worthwhile.

American Samoa: Governor Tanese P.F. Sunia, Marie F. Ma'o, Iotamo Saleapaga, Pita Lauvao, Darryl Cunningham, Etenauga Lutu, Rasela Feliciano, Barbara Ueligitone, Patricia Kalasa, Roger Bartels, Salamo Laumoli, Ann Longnecker, Joseph Tufa, Charles McCutcheon, Charles "Mick" McCuddin, Irene Halshem, William Sword, Noa Ma'ae, Siitia Soliai, Matt Tunoa, and Amu Lapati.

Chuuk: Kiosi Aniol, Herliep Nowell, Sixto Howard, Shinobu Poll, Enrida Pillias, Andita Meyshine, Joakim Peter, and Governor Ansit Walter.

Commonwealth of the Northern Mariana Islands: Isamu Abraham, Josephine Sablon, Greg Calvo, Margaret Aldan, Jon Bruss, Joseph Villagomez, Sr., Joseph Villagomez, Jr., Dr. Sullivan, plus several other teams of staff members of the Department of Public Health, Malua Peter, Heinz Hofschneider, Delores Clark, Gary Bradley, and the irreplacable Bernie Alepuyo.

Fiji: Jimione Samisoni, Wame Baravilala, Joseph Tuisuva, May Okihiro, several undergraduate and graduate students (including several PBMOTP graduates) from Micronesia and Samoa, Mike O'Leary, and Pakawan O'Leary.

Guam: Dennis Rodriguez, PeterJohn Camacho, Marie Borja, many other staff of Department of Public Health and Social Services, Elena Scraggs, the FHP Medical Group, Peter Leon Guerrero, Don Davis, Mary Mantanane, Tyrone Taitano, many staff at Guam Memorial Hospital, Maureen Fochtman, Ulla-Katarina Craig, Augusta Rengiil, Guam Senators Edwardo Cruz and Lou Leon Guerrero, Laurie Duenas, Vince Leon Guerrero, Mary Sanchez, Helen Ripple, Rosario Nededog, Helen Santos, Bernadette Stern, Elizabeth Gray, Barbara Ashe, and Steve and Arlene Cohen.

Hawaii: Sherrel Hammar, Raul Rudoy, Donald Person, Christina Hija, Neal Palafox, Robert Bath, and Gayle Gilbert.

Kosrae: Asher Asher, Hitheo Shrew, Josiah Saimon, several administrators of the Health Department, representatives from the Women's Organization and Women's Affairs, Henry Robert, and Kalwin Kephas.

Republic of the Marshall Islands: Tom Kijiner, Donald Capelle and staff, Zacharias Zacharias, Marie Lanwi, Tom Jack, Oscar de Brum, Bill Graham, Philip Okney, Jack Jorban, Russell Edwards, Cent Langidrik and staff, Alfred Capelle, Troy Barker, Andrew Kuniyuki, the physician staff of Majuro Hospital, Marita Edwin, Norman Smith, C. Eric Lindborg, the mayor of Kwajalein and two councilmen, and the physicians and primary care team at Ebeye Hospital (with special thanks to Dr. Tallens).

Republic of Palau: Masao Ueda, Steve Umetaro, Era Kelulau, Rebecca Bedangel, Ginny Nakamura, Gail Ngirmidol, Regina Mesebeluu, Joanna Polloi, Francisca Blailes, Miriam Chin, Engracia Ongino, Sen. Sandra Pieantozzi, Sen. Santos Olikong, Lucio Ngivaiwet, Gerdence Meyar, Cynthia Malsol, Caleb Otto, Judy Otto, Dr. Ueki, and Francis Matsutaro.

Pohnpei: Eliuel Pretrick, Sizue Yoma, Seamo Norman, Greg Dever, Jan Pryor, Susan J. Moses, Dr. Isaac, and Jacob Nena.

ACKNOWLEDGMENTS

Yap: John Gilmatam, Matthias Kuor, Mathew Haleyaluw, Andrew Yatilman, David Rutstein, Don Evans, Peter Suwei, Anna Boliy, Stan Gufsag, Bayad Untun, and Joe Haleyalmang.

In each jurisidiction, committee members held special meetings with the graduates of the Pacific Basin Medical Officers Training Program. We are extremely grateful to all of them for their willingness to talk about their experiences and were impressed by the dedication each one brought to their duties.

Staff members from several embassies and congressional offices also provided invaluable assistance in providing background information and logistical help before and during the site visits. We extend special thanks to Holly Barker, Samson Pretrick, Susan Yakutis, and Rhinehart Silas.

Several other individuals also provided valuable background information and keen insight into the issues facing residents of the U.S.-Associated Pacific Basin: Nancy Glass, Mac Marshall, Wes Youngberg, and Ralph Rack.

Arranging travel for 13 individuals going to a dozen different islands was truly a logistical nightmare. Our hats go off to Anthony Mavrogiannis and Christina Ieronimo of National Academy Travel.

This project started in 1990 with a planning meeting held in California that highlighted some of the more pressing health issues within the U.S.-Associated Pacific Basin. Polly Harrison, a former IOM staff member, was the program officer for that meeting, and, in the ensuing years, she steadfastly rallied support for this study. Chris Howson, director of IOM's Board on International Health, also provided wise and timely advice once the project actually began. Heather Callahan, project assistant, cheerfully coordinated the complex details of the project—from arranging meetings to making sure that everyone received their reimbursements to formatting the text of the report.

A special thank you goes to Annice Hirt, the project's outstanding research assistant, who weathered Typhoon Isa and remained unshaken by an earthquake measuring more than 6 points on the Richter Scale while on her site visit to Guam. More important, throughout the project she showed tremendous dedication to getting the facts right and to researching thoroughly the historical, social, legal, and economic contexts of the health situation in the Pacific Basin. The project and this report have been greatly enriched by her efforts.

Finally, the artwork on the front cover is by Valerie Hunton, a renowned artist currently living in Fiji. She was named the Honorary Resident Artist of the PBMOTP in Pohnpei, while her husband was on the faculty there. Her works brighten hospitals and clinics throughout the region as well as the covers of the *Pacific Health Dialog*. The work on the cover of this book, *Women and Children*, is from her recently published book, *Pacific Journey: A Celebration.*

Contents

SUMMARY 1

1 **INTRODUCTION** 15
   Origins of this Report, 16
   Study Approach, 16
   Organization of this Report, 17
   U.S. Involvement with the Region, 17

2 **REGIONAL HEALTH AND HEALTH CARE SERVICES
   OVERVIEW** 23
   Regional Health Overview, 23
   Regional Health Care Services Overview, 34
   Appendix, 49

3 **CHARTING A COURSE FOR THE 21st CENTURY:
   A STRATEGIC PLAN FOR FUTURE HEALTH INITIATIVES
   IN THE U.S.-ASSOCIATED PACIFIC BASIN** 55
   Adopt a Viable System of Community-Based Primary Care
     and Preventive Services, 56
   Improve Coordination Within and Between the Jurisdictions and
     the United States, 60
   Role of the United States, 61
   Role of the Island Jurisdictions, 65
   Interface Between the United States and the Island Jurisdictions, 65
   Increase Community Involvement and Investment in Health Care, 67
   Promote the Education and Training of the Health Care Workforce, 71

BIBLIOGRAPHY 77

APPENDIXES
A   Committee Biographies, 85
B   Saipan Workshop Agenda, 89
C   Organization of Institute of Medicine Site Visits to the
    Pacific Basin, 1997, 93
D   Assessments of Individual Jurisdictions' Health Care Services, 95
      American Samoa, 96
      Commonwealth of the Northern Mariana Islands, 104
      Federated States of Micronesia, 110
        Chuuk, 116
        Kosrae, 119
        Pohnpei, 122
        Yap, 125
      Guam (Guahan), 128
      Republic of the Marshall Islands, 141
      Palau (Belau), 149

Tables and Figures

TABLES

2.1 Total Population, U.S.-Associated Pacific Basin Jurisdictions and the United States, 25
2.2 Leading Causes of Death (number of deaths), U.S.-Associated Pacific Basin Jurisdictions, 31
2.3 Cost Summary for DHHS Programs and Activities in the U.S.-Associated Pacific Basin Jurisdictions, by Agency, FY 1995 (in thousands), 42
2.4 Health Resources and Services Administration Budget for Activities in the U.S.-Associated Pacific Basin Jurisdictions, FY 1996, 43
2.5 Health Care Workforce in the U.S.-Associated Pacific Basin, 46
2.6 Number of Physician Graduates, by Year of Graduation, 52
2.7 PBMOTP Budget, FY 1986 to FY 1996, 52
2.8 Number of PBMOTP Graduates Currently Working in U.S.-Associated Pacific Basin Jurisdictions, 53

3.1 Total Health Budget Per Capita, U.S.-Associated Pacific Basin Jurisdictions, 56

FIGURES

2.1 Population estimates for U.S.-Associated Pacific Basin jurisdictions, selected years, 1950–1997, and projected, 1998–2010, 26
2.2 Life expectancy at birth, U.S.-Associated Pacific Basin jurisdictions and the United States, 1996, 27
2.3 Number of infant deaths per 1,000 live births, U.S.-Associated Pacific Basin jurisdictions and the United States, 1996, 28

2.4 Total fertility rate per woman for the U.S.-Associated Pacific Basin jurisdictions and the United States, 1997, 28
2.5 Median ages for populations in the U.S.-Associated Pacific Basin jurisdictions and in the United States, 30
2.6 Percentage of children younger than 2 years of age who are fully immunized, U.S.-Associated Pacific Basin jurisdictions, 1996, 34

3.1 Overview of recommended organizational arrangements, 61

Pacific Partnerships for Health

Charting a Course for the 21st Century

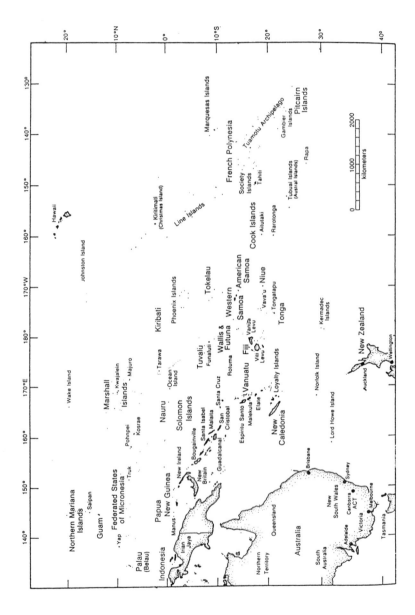

The Pacific region. SOURCE: Pacific Health Dialog: Pacific Peoples of New Zealand. *Journal of Community Health and Clinical Medicine for the Pacific* 4(2), 1997. Copyright 1997 by Resource Books Ltd. Reprinted with permission.

Summary

The U.S.-Associated Pacific Basin consists of six island jurisdictions. Three—American Samoa, Commonwealth of the Northern Mariana Islands (CNMI), and Guam—are considered U.S. flag territories. The other three—Federated States of Micronesia (FSM), Republic of the Marshall Islands (RMI), and Republic of Palau (Palau)—are independent countries, but are freely associated with the United States. The total population of all these jurisdictions is 454,118, roughly the same population as Portland, Oregon, but that population is scattered across 104 inhabited islands covering an expanse of ocean larger than the continental United States. The ties that bind these Pacific islands to the United States have been forged largely within the past century—from ties through trade and religious missions to ties as a result of the United States being the United Nations-approved trust administrator and the preeminent funder of most of the region's economic, social, and health services. The U.S. Department of Health and Human Services provided approximately $70 million in funding for health care in 1996.

The health care delivery systems of the different jurisdictions in the region reflect the challenges and strengths unique to the islands. The health status of the islanders naturally varies within and among the jurisdictions. In general, however, almost all health indicators for islanders are worse than those for mainland Americans. This is most notably so in the freely associated states. The systems must deal with health conditions typical of those of both developing countries (e.g., malnutrition, tuberculosis, dengue fever, and cholera) and developed countries (e.g., diabetes, heart disease, and cancer).

In the delivery of health care services numerous challenges must be overcome. These include: administrative structures that emphasize hospital-based acute care; the long distances that must be covered to provide care to people in remote areas; dependence on foreign aid; inadequate fiscal and personnel management systems; poorly maintained and equipped health care facilities; the enormous costs involved with sending patients off-island for tertiary or specialized care; and shortages of adequately trained health care personnel. In many cases, the island jurisdictions are also contending with significant social change brought about by incredible population growth, rapid economic development, and a shift away from a way of life based on communal farming and fishing to one that is market and consumer oriented. Attempts to address these health conditions and challenges come at a time when U.S. federal government aid to the region has begun to decrease, a trend that is likely to continue.

These challenges are also embedded in the islands' many strengths and resources: cultures that remain vibrant even after years of foreign occupation and influence, strong familial ties and roles for women, highly developed and organized communities, traditional health practices, and powerful religious beliefs.

THE INSTITUTE OF MEDICINE STUDY

The Institute of Medicine (IOM) was asked by the U.S. Department of Health and Human Services, Health Resources and Services Administration (HRSA), to examine these issues and suggest possible approaches to improve the situation. Specifically, IOM was to:

1. collect and examine all readily available information on the status of health service programs in the Pacific Basin, including the recently concluded Pacific Basin Medical Officers' Training Program (described in Chapter 2 and Appendix D);
2. develop assumptions about benchmarks to be used to assess needs and services in those jurisdictions (discussed in Chapter 3);
3. assess the accomplishments, adequacy, and shortcomings of health services programs and related health interventions compared to the baseline assumptions (described in Chapter 2 and Appendix D); and
4. develop a strategic plan to address the problems and inadequacies, and reinforce the successes, in health services (described in Chapter 3).

CURRENT STATUS OF THE HEALTH CARE DELIVERY SYSTEMS

The status of the health care delivery systems varies markedly from island to island. The following is a brief overview of the current situation in each jurisdiction (more complete assessments are provided in Appendix D).

American Samoa: Government management and financial difficulties caused by unpaid debts threaten the functioning of the health care system. Critical supplies have fallen short, the Health Care Financing Administration (HCFA) has threatened decertification of the only hospital, and some basic public health needs are being neglected.

Commonwealth of the Northern Mariana Islands: With a strong legislature and good leadership in the health sector, the CNMI has made significant strides in its health care system. Private insurance markets are beginning to be established, government financing of off-island referrals has been restricted, and the residents are proud of their hospital. However, rapid economic development and the resulting increase in the number of immigrants, primarily from Southeast Asia, brought in for contract work have placed significant strains on the infrastructure and health services.

Federated States of Micronesia: Although the quality of the health care system varies markedly from one state to another among the states of the FSM almost all health care services are extremely dependent on U.S. aid. As the Compacts of Free Association wind down, the resulting economic uncertainty has already begun to restrict some health services.

- **Chuuk:** The most populated and poorest of the four states, Chuuk suffers from a poorly developed infrastructure, crowded conditions, and a seriously inadequate hospital.
- **Kosrae:** Kosrae consists of only one island and has the smallest population of the four states of FSM. It has a relatively good health care system, with positive improvements in public health being seen and acute-care services being provided by a largely indigenous health care workforce. Limited supplies and outdated equipment at health care facilities have been a concern.
- **Pohnpei:** Pohnpei is the home of the former Pacific Basin Medical Officers Training Program and capital of the FSM. It has witnessed significant improvements in its health care systems, especially with the reinvigoration of its primary care system. However, Pohnpei does contend with environmental health concerns associated with poor sanitation and a poorly maintained acute-care facility.
- **Yap:** In general, Yapese are considered the healthiest among the populations of the four FSM states. Even the most populated island of Yap does not have the crowded conditions of the other capital state centers. With strong traditional leadership and community involvement, the Yapese built what was considered a model integrated primary health care system through a network of 30 dispensaries. Today, however, there are concerns that the lack of community support and the lack of government attention to the system are causing it to wither away.

Guam: By comparison with the other U.S.-Associated Pacific Basin jurisdictions, Guam ranks above the other island states in overall wealth and health. It has a well-integrated private insurance market, and a well-developed infrastructure. However, the island still lags behind U.S. mainland states, Alaska, and Hawaii on most health indicators, and is experiencing population growth from neighboring freely associated states that is beginning to strain public health services.

Republic of the Marshall Islands: The RMI has one of the world's highest population densities in its two main urban areas and the youngest population of the six jurisdictions (half of the population is under 16 years of age). Crowded conditions, underdeveloped infrastructure, and poor hospital facilities are just some of the pressures on the health care system. Underlying these conditions is a deep mistrust of the United States, rooted in years of nuclear weapons testing in the area and subsequent research to study the health effects of radiation exposure.

Republic of Palau: With Compact money well secured until 2009 and beyond and a mini-economic boom, Palau is in a financially secure position. A new hospital, a newly trained cadre of Palauan medical officers, and a commitment to improve primary care have boosted the health care system of Palau.

CHARTING A COURSE FOR THE 21ST CENTURY

The committee strongly recommends continued U.S. involvement and investment in the region's health care systems. The nature and scope of this involvement and investment, however, must change. Beyond merely providing health care—a great challenge in its own right—the United States and the island communities must work together with a renewed sense of partnership to produce improved health of Pacific Islanders. Implementation of some of the approaches recommended below has already begun in the jurisdictions. The committee underscores the importance of working on several approaches at the same time. Taken individually the approaches will not have the same impact as the collective efforts and opportunities for potential synergy will be lost.

To achieve this goal of improved health the committee recommends a four-pronged approach:

1. Adopt and support a viable system of community-based primary care and preventive services.
2. Improve coordination within and between the jurisdictions and the United States.
3. Increase community involvement and investment in health care.
4. Promote the education and training of the health care workforce.

The committee believes that priority should be given to the first two goals—adopting a viable system of community-based primary care and preventive services and improving coordination within and between the

SUMMARY   5

jurisdictions and the United States. To a certain extent the other two goals of increasing community involvement and strengthening the health care workforce flow naturally from these first two priorities. The committee cautions, however, that if the two main priorities are not given serious attention by all parties involved, the desired goal of improving the health care systems in the region and ultimately the health of the island populations will not be achieved—even if the other recommendations are fully implemented.

Although it was beyond the charge, expertise, and capability of the committee to make detailed estimations of the costs of implementing these recommendations, the committee believes most of the costs can be covered through the reallocation of current levels of health care funding—especially as a more locally sustainable and viable system of community-based primary care and preventive services is adopted. In the 1993 landmark report, *Investing in Health*, the World Bank calculated the cost of providing a minimum package of public health and essential clinical services in middle-income countries (which all the U.S.-Associated Pacific Basin jurisdictions are considered) in 1990 was $22 per capita. All the jurisdictions currently have considerably higher health budgets per capita—ranging from a high of $614 per capita in the CNMI to a low of $92 per capita in Chuuk (PIHOA, 1997).

Adopt and Support a Viable System of Community-Based Primary Care and Preventive Services

Improve the Island's Physical Infrastructure

Ensuring potable water supplies, adequate sanitation, and reliable electricity throughout the jurisdictions must be a clear priority for both the agencies of the U.S. federal government providing aid and the communities striving to create healthy islands.

Invest in Preventive and Primary Care and Public Health

The committee underscores the vital importance of investment in preventive and primary care and in population-based public health care in the region. Currently, almost all of these activities are funded entirely by the U.S. government. The committee believes it is important to have local governments adequately fund these functions and to provide more funds through local sources. This investment does not necessarily require a great deal of new funding. What it will require, however, is the reallocation of existing funds and the reorganization of delivery systems to more closely integrate acute and primary care systems—a difficult process in any health care system. To combat the inappropriate and discretionary use of funds for health care services, the committee recommends that each jurisdiction place the funds reserved for these purposes in a separate cost center within the overall health care budget.

Maintain or Establish Basic Health Care Standards

The flag territories should maintain HCFA standards. Modification of standards can be considered when a jurisdiction provides appropriate justification. Each freely associated state is strongly urged to establish its own health care standards, including provisions for the licensure of health care providers and legislative practice acts.

Develop a Regional Health Information System

Each of the jurisdictions is encouraged to participate in the development of a standard regional health information system. This system would be the repository of information needed to analyze the health care system and to make assessments of how best to proceed. The system should be able to track outcomes and progress toward *Healthy People 2010* goals that the jurisdictions have revised to capture more appropriately the unique circumstances and disease burdens of their populations.

Reform Health Care Facility Management

Officials at the hospitals and other health care facilities should be given greater control over finances. Each of the jurisdictional governments must also be required to establish an annual budget with separate cost centers for health services facility maintenance and repair, equipment and supplies, salaries, and in-service training. This budget should be tied to an itemized annual work plan. Financial assistance from U.S. sources for facility and equipment repair should be tied to preparation and completion of these annual work plans, to the inclusion of partial financing with local funds.

Promote Prudent Privatization

As many of the jurisdictions have begun or are considering contracting out and privatizing many of their health care services, the committee cautions that technical assistance should be provided to the prospective private businesses. Business plans for these businesses should be developed that include realistic fee collection goals and carefully considered policies regarding supervision and other personnel matters.

Rethink and Restrict Off-Island Tertiary Care Referrals

While recognizing the significant attempts to reduce the enormous costs of such referrals, each jurisdictional government is encouraged to move away rapidly from providing financial support for such referrals altogether and to consider the development of insurance systems, possibly private-market insurance, or other funding mechanisms to cover such catastrophic health care costs.

Improve Coordination Within and Among the Jurisdictions and the United States

To maximize scarce resources and minimize wasteful duplication of efforts, the committee calls for greater coordination and collaboration as well as improved management on both sides of the Pacific. Such coordination has begun, but it must be more focused and more consistently pursued and supported.

Role of the United States

Use of Block Grants That Require Meaningful Measures of Accountability Federal agencies are encouraged to use block grants or to consolidate grants whenever feasible. The committee particularly encourages the use of common grant applications and consolidated reporting formats. The emphasis should be on achieving greater flexibility and efficiency with well-defined measures of accountability and common data systems.

Consider Multiple Uses of Military Facilities The committee recommends that officials from all U.S. military health care facilities in the region enter into dialogue with the jurisdictions to determine the optimal ways of sharing regional resources and providing training opportunities to serve local populations.

Continue to Fund Research Projects in the Region The committee encourages greater interagency coordination to support research and monitoring in the jurisdictions of (1) major causes of morbidity and mortality (e.g., diabetes, substance abuse, tuberculosis, and nutritional deficiencies); (2) health systems development; and (3) health effects from radiation exposure throughout the Pacific Basin jurisdictions, albeit with continued specific attention to the Marshall Islands. The conduct of the research must be in accordance with accepted ethical principles and with the full cooperation of the island communities being studied.

Establish an Interagency Governmental Committee on Pacific Health The committee believes that coordination of U.S. funding for all health-related activities in the Pacific is needed to increase the coherent and consistent application of rules, regulations, and accountability requirements for expenditures, which should be based on the previously discussed outcomes measures. The committee therefore recommends the establishment of an Interagency Governmental Committee on Pacific Health (IGCOPH) to ensure coordination of health programs, health program administration, and technical assistance to the region. The committee should be chaired by the Secretary of the U.S. Department of Health and Human Services or his or her designee and include representation from each of the federal agencies that fund health-related activities in the region, including, but not limited to the following:

- U.S. Department of Health and Human Services (Administration on Children, Youth, and Families; HCFA; HRSA; Centers for Disease Control and Prevention; and the Substance Abuse and Mental Health Services Administration);
- U.S. Department of the Interior (Office of Insular Affairs),
- U.S. Department of Agriculture,
- U.S. Department of Commerce,
- U.S. Department of Education,
- U.S. Department of Energy,
- U.S. Department of Defense, and
- U.S. Department of State.

Roles of the Island Jurisdictions

The island jurisdictions are also encouraged to continue their commitment to and collaboration in the Pacific Island Health Officers' Association (PIHOA). PIHOA is further encouraged to: (1) develop a regional health information system to promote a shared version with standard nomenclature, (2) continue to review purchasing practices and encourage shared purchasing and volume buying to decrease costs and to be able to share resources in emergencies, and (3) continue to identify technical assistance and consulting strategies that promote the prudent use of the expertise available within the region.

Interface Between the United States and the Island Jurisdictions

Establish a Pacific Basin Health Coordinating Council. Finally, the committee recommends that the governments of the United States and the six island jurisdictions establish or designate a nongovernmental organization in the region to coordinate health affairs and facilitate collaboration between the U.S. and jurisdiction governments. This "Pacific Basin Health Coordinating Council," or PBHCC, would meet quarterly and report annually on the progress of health sector reform in the U.S.-Associated Pacific Basin to the President of the United

SUMMARY 9

States, U.S. Congress, the chief executive officers and legislatures of each island jurisdiction, IGCOPH, and PIHOA. The PBHCC should have a small permanent staff. The establishment of such a Council is not meant to create yet another layer of bureaucracy; rather, it is envisioned as the catalyst for pragmatic health reforms and the watchdog for greater accountability of all parties—in the United States and the region.

PBHCC Composition. The 14-member Council should have representation from three different groups: 4 representatives of the U.S. government and the IGCOPH; a representative of each island jurisdiction's government; and a total of 4 private citizens, 2 each from the United States and the island jurisdictions.

PBHCC Tasks. What projects are undertaken by the PBHCC will need to be determined cooperatively with all the parties involved. The committee notes the differences in budgets, health care services, personnel, and program directions between the U.S. flag territories (American Samoa, CNMI, and Guam) and the freely associated states (FSM, RMI, and Palau). Therefore, as it undertakes the following potential tasks, the PBHCC should consider grouping jurisdictions accordingly. Specific PBHCC tasks could include:

- help develop health care priorities in the six jurisdictions that take into account burden of disease and cost-efficiency criteria, as well as community-developed priorities;
- ensure coordination of U.S. health programs, activities, and funding streams within the U.S. flag territories and the freely associated states;
- review each jurisdiction's progress toward meeting specific health outcome objectives as required under various U.S. grants;
- assist with the simplification of filing and reporting formats and forms for various U.S. grants;
- facilitate the provision of technical assistance and training in such areas as health administration for all Ministers and/or Directors of Health and for members of jurisdictional Health Authorities or Boards; and
- identify and establish working relationships with U.S. federal agencies and international organizations and other aid donors (e.g., the World Health Organization, the Asian Development Bank, the South Pacific Commission, and others) who could provide technical assistance resources for health care reform.

PBHCC Funding. Funding for this nongovernmental organization must come from a variety of sources. As described in Chapter 1, the U.S. federal government has several vital interests in investing in the region's health and ensuring that the money it provides is spent wisely. In keeping with several of its other recommendations, the committee underscores its belief in the vital importance of having the local jurisdictions provide financial support to this endeavor. The committee also sees a role for private organizations and foundations—both inside and outside the region—to play in funding the

PBHCC. All these funding partners must believe that they have a stake in the PBHCC's work and will benefit from the results that work produces.

Several possible funding mechanisms exist; the committee suggests a few options here, but ultimately the various funders must collectively determine exactly how each will pay. The U.S. federal government might consider contributing a fixed percentage of all funds it provides to the region. Similarly, each of the island jurisdictions may decide to base its funding on a fixed percentage of its total health care budget or on a fixed percentage of the total funds it receives from the U.S. government and other international sources.

Increase Community Involvement and Investment in Health Care

The committee believes that any attempt to improve health care in the Pacific Basin must tap into the strengths and resources of the community—if the improvements are to be meaningful and sustained. Fostering an environment that enables households to improve the health of their members, particularly by promoting the rights and status of women, is viewed as an essential precondition for improving global health. This focus on women is particularly apt in the Pacific, given the central roles of women and girls in many of the island societies.

The committee acknowledges the differences in institutional capacities of each of the jurisdictions and in the cultural norms and functioning of individual communities. No one paradigm of community involvement applies to all island cultures equally or necessarily appropriately. Health services must be aligned to each community's needs and must be congruent with its unique culture, with special attention given to the most vulnerable groups. Each community will have to determine how best to achieve the levels of involvement and investment needed to truly make a difference in the health of its population.

The following are some fundamental steps that island communities should consider.

Establish Jurisdictional Health Authorities

Where appropriate, individual jurisdictions should create, through local legislation and community input, an independent authority or board to oversee the administration of the health care system, plan and prioritize health initiatives, and provide accountability. Such an authority or board would oversee the budgets of all the health services, agencies, hospitals, primary care sites (dispensaries), and programs under its direction through the development of sound annual budget development practices, the use of monitoring systems, and timely annual audits. The Health Authority or Board should have both men and women, and include community volunteers such as business people, clergy, educators, and health care professionals.

Develop Health Improvement Benchmarking Process

Individual jurisdictions and communities should establish a process for determining what health issues are of greatest concern, how best to address those concerns, and how they will monitor their progress. This belief in the power of community involvement and decisionmaking, rather than outside consultants evaluating the current situation, lead the committee to decide not to revise or modify the existing benchmarks used by the University of Hawaii in two evaluations conducted during the 1980s.

Use Nongovernmental Community Organizations to Provide Health Care Services

Nongovernmental community organizations—particularly women's, church, and youth groups—represent a potent and much underutilized force. Such groups should be enlisted to provide a variety of health-related activities, including health education and treatment (e.g., drug counseling), whenever possible.

Increase Community Involvement with Primary Care Sites

For all jurisdictions, when U.S. funds are involved, community commitment and involvement in the delivery of care and the maintenance of primary care sites (dispensaries) should be required. Minimal requirements would be (1) donation of land from the community or some of its members, with a clear deed attesting to the donation; (2) contributions to the construction of the facility in the form of either materials or labor; (3) commitment of the community to maintaining the facilities; and (4) contribution to the salary of the person(s) serving as community health aide(s). In-kind donations should consist of time and labor as well as or in place of monetary contributions.

Promote Education and Training of the Health Care Workforce

The committee is gravely concerned about maintaining the skills and knowledge of the current health care workforce and strengthening the region's local human capacity. The committee therefore recommends several education activities for health professionals to address the present lack of adequate training opportunities available to the health care workforce in the U.S.-Associated Pacific Basin. *The exact nature of these activities, however, should be based on a workforce development and training plan established by each jurisdiction.* The plan should consider not only how to enhance and improve the skills of current health care providers but also how to train new providers, particularly women, to address shortages and natural loss through retirement and attrition.

Activities should include, but should not be limited to the following.

Improve and Support Basic Education

Currently, the primary and secondary educational systems throughout the region do not adequately provide students the skills that they need to participate in the health care workforce. In the short term, targeting special programs to students interested in and academically able to pursue careers in health care should be considered. Public and private scholarships for health care education should also be promoted.

Utilize Distance-Based Learning, Telemedicine, and Electronic Data Libraries

Given the remoteness of the islands, strategies that maximize the efficient and affordable use of these technologies need to be supported.

Provide Postgraduate and Continuing Medical Education Programs

Continuing medical education (CME) is currently provided in a rather haphazard fashion, if at all. CME must be required for all levels of practitioners and incorporated into each jurisdiction's health care workforce training plan. The committee is particularly concerned about the graduates of PBMOTP and recommends that they receive continuing medical education to improve and maintain their clinical skills and knowledge. Advanced training should be conducted at a regional training center, preferably an existing one. The role of the U.S. government should be directed toward capacity-building and financial assistance.

Sponsor Training for Dentists

The committee is greatly concerned by the dearth of dentists currently practicing in the region. Compounding this problem is the fact that many of the current dental practitioners are expatriates or are nearing retirement. The U.S. federal government and local jurisdictions should sponsor dental training immediately.

Sponsor Training for Nurses

Nurse training in the region also needs to be reinvigorated. The committee believes it should continue to take place, as it does now, in several institutions of higher education located throughout the region. The nurse training programs throughout the region are encouraged to work together. This could include the sharing of faculty members, using cooperative efforts to provide distance-based education, upgrading of the curriculum to the bachelor's level, and the development of continuing nursing education programs for existing nurse personnel of all levels.

Provide Health Administration and Systems Management Training to the Chief Health Administrator

Advanced training for the jurisdiction's chief health administrator should be provided through a certificate or degree program. The training should include coursework taken at institutions of higher education that offer high-quality programs in this subject area combined with practical applications and fieldwork within the administrator's own jurisdiction.

1

Introduction

Covering an expanse of ocean larger than the continental United States, the U.S.-Associated Pacific Basin consists of six island jurisdictions. Three are U.S. flag territories: American Samoa, the Commonwealth of the Northern Mariana Islands (CNMI), and Guam. The other three—Federated States of Micronesia (FSM), Republic of the Marshall Islands (RMI), and Republic of Palau—are independent countries but are also freely associated with the United States. This report examines one aspect of U.S. involvement: its role in the region's health care delivery systems.

The health care delivery systems of the different jurisdictions in the region reflect the challenges and strengths unique to the islands. The health status of the island inhabitants naturally varies within and among the jurisdictions. In general, however, almost all health indicators for these islanders are worse than those for mainland Americans. This is most notably so in the freely associated states. The systems must deal with health conditions typical of those of both developing countries (e.g., malnutrition, tuberculosis, dental caries, dengue fever, and cholera) and developed countries (e.g., diabetes, heart disease, and cancer).

In the delivery of health care services numerous challenges must be overcome. These include: administrative structures that emphasize hospital-based acute care; the long distances that must be covered to provide care to people in remote areas; dependence on foreign aid; inadequate fiscal and personnel management systems; poorly maintained and equipped health care facilities; the enormous costs involved with sending patients off-island for tertiary or specialized care; and shortages of adequately trained health care personnel. In many cases, the island jurisdictions are also contending with significant social change brought about by incredible population growth, rapid economic development, and a shift away from a way of life based on communal farming and fishing to one that is market and consumer oriented. Attempts to address these health conditions and challenges come at a time when U.S. federal government aid to the region has begun to decrease, a trend that is likely to continue.

These challenges are also embedded in the islands' many strengths and resources: cultures that remain vibrant even after years of foreign occupation and influence, strong familial ties and roles for women, highly developed and organized communities, traditional health practices, and powerful religious beliefs.

ORIGINS OF THIS REPORT

The Institute of Medicine (IOM) was asked by the Health Resources and Services Administration (HRSA) of the U.S. Department of Health and Human Services to examine these issues and suggest possible approaches to improve the health care situation. Specifically, IOM was to:

1. collect and examine all readily available information on the status of health service programs in the Pacific Basin, including the recently concluded Pacific Basin Medical Officers' Training Program (described in Chapter 2 and Appendix D);
2. develop assumptions about benchmarks to be used to assess needs and services in those jurisdictions (discussed in Chapter 3);
3. assess the accomplishments, adequacy, and shortcomings of health services programs and related health interventions compared to the baseline assumptions (described in Chapter 2 and Appendix D); and
4. develop a strategic plan to address the problems and inadequacies, and reinforce the successes, in health services (described in Chapter 3).

Additional financial support for this project was provided by the Office of Insular Affairs of the U.S. Department of the Interior and the Archstone Foundation.

STUDY APPROACH

To undertake the requested study, IOM's Division of Health Care Services and Board on International Health convened a 12-member study committee, with experts in primary health care, education, international health, mental health, and public health (see Appendix A for a complete list of committee members and their brief biographies). The committee met three times between January and August 1997. The committee's second meeting occurred in April 1997 on Saipan, CNMI, in conjunction with a workshop with health officers from the region (see Appendix B for the workshop agenda and a list of

INTRODUCTION

participants). Immediately before the Saipan workshop, teams of committee members visited all the jurisdictions (see Appendix C for more details).[1]

Committee and staff reviewed the literature on health care services in the region and met with the jurisdictions' key congressional and diplomatic representatives in Washington, D.C. Grant applications and reports on federal grants were also reviewed. Two papers were commissioned: one examined the social and cultural implications of health care delivery in the region, and the other provided updated health data for the region. At its third and final meeting in August 1997, the committee reviewed a draft manuscript and discussed final conclusions and recommendations.

ORGANIZATION OF THIS REPORT

This document constitutes the committee's final and formal report. It is divided into three main chapters. This chapter provides background on the IOM's study and U.S. involvement with the region's health care. Chapter 2 presents an overview of the region's health and health care services, including a chapter appendix that describes the Pacific Basin Medical Officers' Training Program. Finally, Chapter 3 contains the key recommendations for a plan for future health initiatives in the region. General background information and assessments of each of the six jurisdiction's health care delivery systems can be found in Appendix D.

U.S. INVOLVEMENT WITH THE REGION

The ties that bind these Pacific islands to the United States have been forged largely within the past century—from ties through trade and religious missions to ties with the United States as the United Nations-approved trust administrator and the preeminent funder of much of the region's economic activity and its social and health services. As the twentieth century ends, these ties continue to change. Yet, the United States still has a variety of important military, economic, and health interests in the region.

Historical Overview

Archaeological data suggest that the islands of the U.S.-Associated Pacific Basin might have been inhabited as far back as 3000 B.C. Contact with the Western world, however, began in 1521, when the Spanish explorer, Ferdinand Magellan, landed in the Mariana Islands. Since that time, the islands have

---

[1]The site visit to RMI took place in July 1997.

experienced successive waves of foreign occupation and domination: Spanish, German, Japanese, and, most recently, American.

Spain remained in control of all the islands except the Marshall Islands and American Samoa until 1898. That is when, shortly after the Spanish-American War, the United States acquired Guam. Germany, through a series of diplomatic maneuvers, purchased many of the remaining islands from Spain. German interest in the region was primarily economic; the interest of the United States was decidedly economic as well, although missionaries had begun working in the region as early as the mid-1850s. In the South Pacific, American Samoa was officially claimed as a U.S. territory in 1900. Fourteen years later, Japan took control of the German-associated Micronesian islands at the start of World War I; its rule continued until the end of World War II. During its occupation, thousands of Japanese citizens immigrated to the islands and helped to develop an extensive infrastructure of roads, schools, and hospitals.

After World War II, the United States reclaimed control over Guam. The United States also took administrative control over all the other Micronesian islands: first through a military administration and later through the United Nations' Trust Territory of the Pacific Islands (TTPI), administered through the U.S. Department of the Interior (DOI). A substantial military presence built up during the Cold War because of the islands' strategic position near Asia and for weapons testing, training maneuvers for the Central Intelligence Agency (CIA), and military staging for the Korean and Vietnam conflicts.

Current Political Status of the U.S.-Associated Pacific Basin Jurisdictions

The political status of most of the U.S.-Associated Pacific Basin jurisdictions has changed greatly over the past 20 years. Currently, the three flag territories (American Samoa, CNMI, and Guam) are officially part of the United States. The three freely associated states (FSM, RMI, and Palau) are independent countries that have chosen to be associated with the United States through Compacts of Free Association.

The Flag Territories

Guam, CNMI, and American Samoa maintain close ties to the United States either as unincorporated territories (Guam and American Samoa) or through a commonwealth covenant (CNMI). All citizens of the flag territories are U.S. citizens. A commonwealth differs from a territory in that the commonwealth can control its own immigration, customs, and tax policies. An unincorporated territory means that not all provisions of the U.S. Constitution apply to the territory. All three flag territories have locally elected legislatures and governors.

The Freely Associated States

The FSM, RMI, and Palau marked their political independence by signing Compacts of Free Association with the United States. Each Compact provides for development assistance and cedes full authority and responsibility for the jurisdiction's defense to the United States. The Compacts also allow citizens of the freely associated states to immigrate to the United States and any of its territories and possessions. FSM and RMI signed 15-year Compacts that will expire in 2001. Renegotiation of the Compacts is allowed, however, and FSM and RMI have apparently chosen to pursue this option (U.S. State Department, 1996). Palau signed a 50-year Compact that will expire in 2044.

U.S. Military Interests in the Region

The United States has had a military presence in the region that can be traced back to 1898, when the United States acquired Guam after the Spanish-American War. From 1946 to 1958, the U.S. Department of Defense performed nuclear weapons testing in the Marshall Islands (U.S. State Department, 1994). Saipan was the CIA's training ground for Chinese Nationalists in preparation for a possible assault on Communist China (Hezel, 1995). In the aftermath of the Cold War, a strong military presence remains a priority for the United States and its partners in Asia. The United States is obligated to protect Japan, Taiwan, and—through the Compacts—the freely associated states from external threats. In addition, the U.S. presence in the Pacific Basin is considered a source of strength in Asian affairs and protects trade routes.

The islands continue to hold a significant geographic importance in the Pacific, particularly as China develops its economic and military power. The United States military has unlimited access to the waterways of all the jurisdictions, as well as the authority to turn away vessels from other nations (Bank of Hawaii, 1996). The past and current instability on the Korean Peninsula has required the presence of U.S. troops in South Korea, but they are supplemented by the troops in the Pacific Basin. This is particularly important because U.S. bases in the Philippines have closed and the number of U.S. forces in Okinawa, Japan, may decrease.

U.S. Economic Interests in the Region

Today, more than 40 federal agencies—from DOI to the U.S. Postal Service—are involved with all six jurisdictions in various ways. The United States also has an interest in the millions of dollars that it has and will continue to provide the region. The three flag territories are officially part of the United States and are therefore eligible for federal economic aid and programs, much like any of the 50 states. Likewise, the United States still remains heavily involved with providing financial support to the freely associated states, and each of the Compacts calls on the United States to promote greater economic self-sufficiency for each of the jurisdictions. Currently, more U.S. dollars are spent per capita on freely associated states than any other foreign country (Arvis, 1997).

Of the millions of dollars spent on the islands, however, a large amount is returned when islanders import U.S. goods (Diaz, 1997). Consumer goods and conveniences, such as Spam® and beer (albeit not the most healthy of trade items) represent major commodities in Pacific trade with the United States (DOI, 1996). Although the region imports far more than it exports, it does export some commodities—ranging from tuna canned in American Samoa, to garments made in CNMI, to gourmet pepper grown in Pohnpei. Copra (dried coconut meat) is still produced in many of the region's outer islands. Tourism is the economic mainstay for many of the islands and this industry is growing throughout the region (DOI, 1996).

U.S. Interests in the Region's Health Care Systems

As the Compacts were being negotiated in the early 1980s, all parties agreed that the United States should remain involved with the region's health care. Today several federal agencies provide health-related services and funding to the region, including several agencies of the U.S. Department of Health and Human Services (DHHS), U.S. Department of Energy, and DOI. For example, in fiscal year 1996, DHHS provided approximately $70 million to the region (see Chapter 2 for more detailed figures) (HRSA, 1996). DOI gave funds to improve the jurisdiction's infrastructure, including some of its health care facilities, and provided technical assistance in a range of areas related to health care. Additionally, all Compact and Covenant funds for the islands are administered through DOI.

The benefits of improved health in the Pacific flow not only to islanders but also to populations on the U.S. mainland and other countries as well (IOM, 1997a). In an age of nearly instant access to every corner of the globe, no one is immune to the threat of infectious diseases. This is of special concern in the Pacific, where poor environmental and sanitary conditions and high degrees of mobility make island populations much more vulnerable than the populations in industrialized nations. As the Pacific Basin jurisdictions grow in population and attempt to modernize, problems persist with basic sanitation and environmental health. In the FSM and the RMI, respiratory and waterborne illnesses are a major concern (Hezel, 1997). The Centers for Disease Control and Prevention (CDC) advises travelers to the region—particularly in the remote outer islands—to take precautions against measles, hepatitis B, parasites, and cholera (CDC, 1996). Although infectious diseases remain an important public health concern, few places in the region can quickly collect and analyze data on these communicable diseases (Diaz, 1997).

The Compacts of Free Association permit the free movement of people between the freely associated states, flag territories, Hawaii, and the mainland United States. As Diaz (1997) points out, "At any given moment, an epidemic can break out in the Pacific that might remain undetected until a laboratory result returns from Hawaii several days later. By then, the highly transient population might have carried the disease to the population centers or a tourist might have carried it back to Hawaii or the U.S. mainland" (p. 121). Recently, an outbreak of measles on Guam spread to Chuuk, despite a campaign to prevent the spread of the disease by Guamanian health officials. In 1991, dengue fever spread from Palau to Yap and then Guam before the epidemic was detected (Diaz, 1997).

The United States has also benefited from health research conducted in the islands. By documenting the health status and treating the Marshallese islanders and the U.S. military personnel exposed to radiation during nuclear weapons tests during the 1940s and 1950s, for example, the United States has learned a great deal about radiation and its effects on human health (Howard et al., 1995). In Guam, research on the neurodegenerative syndromes Lytico and Bodig has yielded valuable information and basic science now being used by scientists to better understand Alzheimer's disease and other forms of dementia (Sacks, 1996).

2

Regional Health and Health Care Services Overview

> *Healthy islands should be places where:*
> *children are nurtured in body and mind;*
> *environments invite learning and leisure;*
> *people work and age with dignity;*
> *ecological balance is a source of pride.*
>
> —*Yanuca Island Declaration on Health in the Pacific in the Twenty-First Century* (WHO, 1995)

The Yanuca Island Declaration represents the collective health goals of all Pacific Basin jurisdictions. How close the islands actually come to achieving this vision varies considerably. This report focuses primarily on the health care delivery systems in the six jurisdictions of the U.S.-Associated Pacific Basin. This chapter begins with a brief overview of some of the region's key health status indicators. They are a reflection, in part, of how effectively health care services are being provided. The second section of this chapter provides an overview of the region's health care services. (Specific information on the health care services in each jurisdiction is provided in Appendix D.) The chapter concludes with an appendix describing the recently concluded Pacific Basin Medical Officers Training Program (PBMOTP) based in Pohnpei, FSM, which trained 70 islanders as physicians.

REGIONAL HEALTH OVERVIEW

Within the past century, the region as a whole and the U.S.-Associated areas in particular have experienced both demographic and epidemiological transitions. The demographic transition means that people are living longer (e.g., they have longer life expectancies and the infant mortality rate is lower). The epidemiological transition means that people tend to die from non-communicable diseases (e.g., heart disease and cancer) rather than infectious diseases (e.g., influenza and tuberculosis). Guam and the Commonwealth of the Northern Mariana Islands (CNMI) appear to have already completed these transitions. Importantly, however, many of the jurisdictions (particularly in the freely associated states) appear to be experiencing both transitions at roughly the same time and must therefore confront the challenges of simultaneously providing health care to older individuals with chronic conditions as well as to younger individuals fighting acute infectious diseases.

Demographic Characteristics of the Region

Population

Roughly half a million people (454,118) live in the six jurisdictions of the U.S.-Associated Pacific Basin (American Samoa, CNMI, Guam, Federated States of Micronesia [FSM], Republic of the Marshall Islands [RMI], and the Republic of Palau [Palau]) (see Table 2.1). Although the islands cover an area of the Pacific Ocean that is larger in size than the continental United States, most people live on a handful of densely populated islands. Currently, Guam, with 155,225 people has the largest population of the six jurisdictions; Palau, with only 17,225 people, has the smallest population (PIHOA, 1997). (For more information on population for each of the jurisdictions, see Appendix D.)

Overall, the region has experienced a high rate of population growth since 1950 (the approximate start of U.S. involvement and administration in all jurisdictions), and that growth is projected to continue to increase rapidly over the next few years (see Figure 2.1). The increase in the total population and projected population growth result from several factors: higher life expectancy (see Figure 2.2); lower infant mortality (see Figure 2.3); and, in some jurisdictions such as CNMI and Guam, high rates of immigration.

One change typical of a completed demographic transition is a decline in fertility rates and a resulting increase in the median age. As indicated in Figures 2.4 and 2.5 respectively, however, several of the jurisdictions have not completed their demographic transition because they continue to have high fertility rates and low median ages. In RMI, for example, the median age is 16.2 years. This means that half of the population of RMI is under the age of 16. This could have a tremendous impact on growth rates as more and more women reach childbearing age.

Migration is another factor that has contributed to population changes in the jurisdictions, as well as in the Pacific region in general. In search of better economic opportunities, Pacific islanders have migrated to Australia, New Zealand, and the United States, including its territories.[2] High rates of

---

[2]According to the 1990 U.S. census, 56,153 Micronesians (49,345 Guamanians and 6,808 "other Micronesian") were residing in the United States (Bureau of the Census, 1990).The census also shows that Samoans numbered 62,964 in 1990. Although not counted in the census, numbers of citizens from the freely associated states living in American Samoa, CNMI, Guam, and Hawaii have been estimated to assess the impact of the Compacts of Free Association on Hawaii and U.S. territories. For a discussion of these population estimates, see Appendix D, under either CNMI or Guam.

immigration—from other Pacific island countries and Southeast Asia—have contributed to significant population increases in some places such as CNMI and Guam. On islands with fewer economic opportunities, large numbers of citizens emigrating out have helped to lower population growth (East-West Center, 1996). For example, an average of about one percent of the FSM population emigrates each year since the Compacts went into effect in 1986, significantly contributing to declining growth rates in that jurisdiction (Hezel, 1997) (See also Appendix D, CNMI and Guam assessments, Compact impact descriptions.)

TABLE 2.1 Total Population, U.S.-Associated Pacific Basin Jurisdictions and the United States, 1997

Jurisdiction	Total Population
American Samoa	58,070
CNMI	58,846
Guam	155,225
FSM	105,506
Chuuk	53,319
Kosrae	7,317
Pohnpei	33,692
Yap	11,178
RMI	59,246
Palau	17,225
TOTAL, Pacific Basin jurisdictions	454,118
TOTAL, United States	260,372,174

NOTES: Total population is the total number of people residing in the jurisdiction. Official estimates for American Samoa, Guam, and Palau are according to the Bureau of the Census 1990 reports. The population for CNMI is from the 1995 mid-decade census. The population for the RMI is from 1996 and that for FSM is from the 1994 FSM national census. Total U.S. population is from July 1, 1994, official estimate based on the 1990 U.S. Census. It includes the resident population living in the 50 U.S. states and the District of Columbia.

SOURCES: Population figures for the six jurisdictions are from PIHOA (1997); the U.S. population is from the Bureau of the Census (1997a).

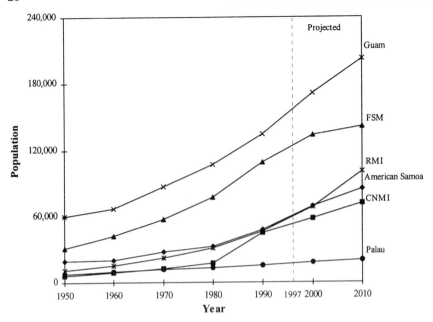

FIGURE 2.1 Population estimates for U.S.-Associated Pacific Basin jurisdictions, selected years, 1950–1997, and projected, 1998–2010. SOURCE: Bureau of the Census (1997b). NOTE: Projections are based on average annual growth rates for each jurisdiction. NOTE: Assumptions for estimates and projections are based on fertility, mortality, and migration assumptions that may differ somewhat from those provided by PIHOA, 1997. For more information, please see Bureau of the Census, International Population Center (URL www.census.gov/ipc).

Ethnicity

Ethnically, most of the population in the North Pacific (CNMI, Guam, Palau, FSM, and RMI) is considered Micronesian. In the South Pacific (American Samoa), the population is almost entirely Polynesian. (More precise breakdowns of each jurisdiction's ethnicity are given in the jurisdiction's individual assessment found in Appendix D.) This general homogeneity of the overall population masks considerable diversity of culture and language—even within the same jurisdiction. For example, in FSM, eight major indigenous languages are spoken and no two states have the same native language (Hezel, 1997). English is spoken throughout the region, but for most people it is a second language.

Economy

Economic conditions vary tremendously throughout the region. In general, residents of the flag territories enjoy a higher standard of living than residents in the freely associated states. For example, in 1994 the per capita gross island product for Guam was $20,640, but in the FSM it was only $2,000 (Bank of Hawaii, 1995a; Bank of Hawaii, 1995b). Economic conditions and characteristics are described in more detail in the individual jurisdictional assessments given in Appendix D.

Epidemiological Characteristics of the Region

The health status of the islanders naturally varies within and among the jurisdictions. In general, however, almost all health indicators for islanders are worse than those for mainland Americans. This is most notably so in the freely associated states. The health care systems must deal with health conditions typical of those of both developed countries (e.g., diabetes, heart disease, and cancer) and developing countries (e.g., malnutrition, tuberculosis, dengue fever, and cholera). Key health promotion and disease prevention indicators are of concern as well.

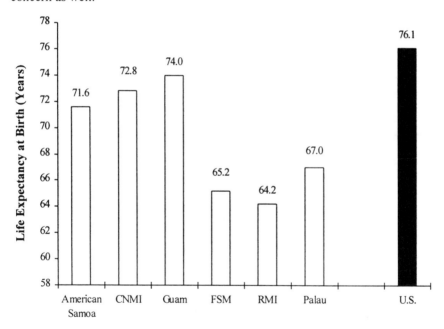

FIGURE 2.2 Life expectancy at birth, U.S.-Associated Pacific Basin jurisdictions and the United States, 1996. SOURCES: Ages for the six jurisdictions are from PIHOA (1997); age for the United States is from NCHS (1997) and represents preliminary 1996 data. NOTES: Life expectancy at birth is defined as the number of years that a person at birth is expected to live under the mortality pattern prevalent in the community or country. Data for the six jurisdictions are from their respective most recent censuses or population surveys: Palau, 1996; Guam and American Samoa, 1990 censuses; CNMI, mid-decade census; FSM, 1994 census; and RMI, 1994 survey.

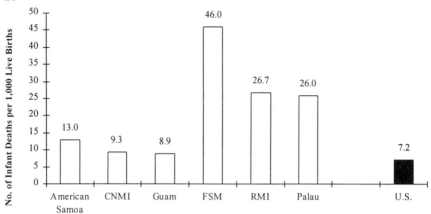

FIGURE 2.3 Number of infant deaths per 1,000 live births, U.S.-Associated Pacific Basin jurisdictions and the United States, 1996. SOURCES: PIHOA (1997); the U.S. rate is from NCHS (1997) and represents preliminary 1996 data; the goal for infant mortality rate set in Healthy People 2000 (USDHHS, 1996) is 7 per 1,000 live births. NOTES: An infant is considered to be a child between the ages of birth and 1 year. RMI reported 63.0 infant deaths per 1,000 live births in 1994.

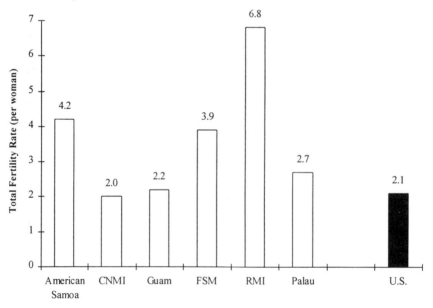

FIGURE 2.4 Total fertility rate per woman for U.S.-Associated Pacific Basin jurisdictions and the United States, 1997. SOURCES: For CNMI, PIHOA (1997); for all others, Bureau of the Census (1997b). NOTE: RMI reported a rate of 5.7 from its mid-decade 1995 census.

Mortality

The leading causes of death in the region (see Table 2.2) are very similar to those of developed countries, with heart disease and diseases of the circulatory system leading the list in all jurisdictions except FSM and RMI. Tobacco use and abuse is very prevalent throughout the region and has been linked to many of the leading causes of death (Marshall, 1991). Accidents, especially those involving motor vehicles, are also a leading cause of death; possible reasons include the densely populated areas where most people live as well as the abuse of alcohol and other substances.

Disease Prevalence

The focus of this report is on the health care delivery system rather than health status. As such, the committee did not make a special effort to collect data on disease rates and prevalence although such information can indicate how well the health care delivery system is working. Unfortunately, such data are not readily or consistently available from all jurisdictions. It is important to note that Pacific Islanders have not generally been included in most *Healthy People 2000* objectives (in fact Epstein reports that out of the hundreds of *Healthy People 2000* objectives, only eight address Asian Americans and Pacific Islanders directly and none address the U.S.-Associated Pacific Basin jurisdictions [1997]). The overall disease trends point to an increase in the prevalence of noncommunicable diseases and a decrease in communicable and infectious disease. The following list is meant to provide a brief overview of some of the more pressing health concerns in the islands. It is by no means comprehensive.

Diabetes Just after World War II, a naval survey of the islands found no cases of diabetes (Flear, 1997b). Today, however, diabetes is a major health concern in each of the jurisdictions. The dramatic increase is associated with many factors, most importantly increased use of fatty or salty imported food, increased consumption of alcohol, and decreased physical activity (Brewis et al., 1996). Diabetic patients contribute to the high demand for dialysis, and the effects of uncontrolled diabetes are major reasons for off-island referrals. In Guam the prevalence of middle-age-onset diabetes is seven times that in the United States. It accounted for about 5.1 percent of all deaths on Guam between 1983 and 1992 (GHPDA, 1996). In RMI, 30 percent of the population over 15 years of age suffers from diabetes (Diaz, 1997).

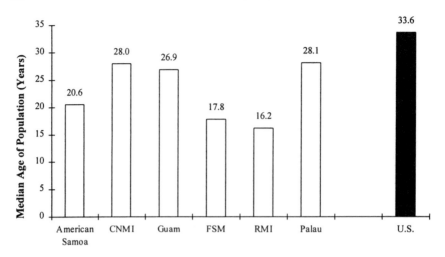

FIGURE 2.5 Median ages for populations in the U.S.-Associated Pacific Basin jurisdictions and in the United States. SOURCE: U.S. Department of the Interior, Office of Insular Affairs (1996). NOTES: Data for the six jurisdictions are based on: 1990 Census reports for the areas, American Samoa census, 1995; CNMI census, 1995; FSM census, 1994; Guam census, 1995; RMI census, 1988; and Palau census, 1994. Median age for the United States is from the Bureau of the Census, *Current Population Survey,* 1995.

Cancer With the exception of FSM, all jurisdictions list cancer as one of the top three causes of death. On Guam, a recent study of death certificates showed between 1971–1995 lung cancer was responsible for a little over one-third of all recorded cancer deaths (Haddock, 1997). In RMI, one survey of Nuclear Claims Tribunal records found that lung and cervical cancers accounted for over two-thirds of cancer incidences in that population between 1985 and 1994. Those cancers, and others, occurred at rates significantly higher than in the United States. Cervical cancer was 5.8 times higher in Marshallese females than in U.S. females. (Palafox, Johnson, Katz, et al., in press). Some evidence has linked certain oral cancers and betel nut chewing, a widespread practice in the Marianas, Yap, and Palau. However, this link is not well-established since studies have not taken into account other risk factors such as the use of tobacco, which is commonly chewed with betel nut (Haddock, 1997; Haddock, et al., 1981; Marshall, 1987, 1991).

Tuberculosis Tuberculosis (TB) is a problem within all the jurisdictions. FSM had 171 registered cases of TB in 1994, with an incidence rate of 163/100,000 (FSM, 1996). Guam, CNMI, and Palau have also noted increases in the numbers of cases of TB and attribute these increases in large part to the arrival of large numbers of foreign contract workers from Southeast Asia, who are more likely to have and spread the disease. Guam, for example, has a TB incidence rate seven times the U.S. rate (Diaz, 1997). They have also encountered drug-resistant TB, and reoccurrence of the disease in older people who were treated in their earlier years.

TABLE 2.2 Leading Causes of Death (number of deaths), U.S.-Associated Pacific Basin Jurisdictions

American Samoa, 1995	CNMI, 1991	Guam, 1995	FSM, 1994	RMI, 1996	Palau, 1996
Heart diseases (50)	Circulatory (36)	Heart diseases (175)	Circulatory (86)	Sepsis and septic shock (29)	Circulatory (43)
Cancer (32)	Accidents (25)	Cancer (100)	Respiratory (53)	Renal failure and disease (16)	Accidents (28)
Cerebrovascular (18)	Cancer (16)	Accidents (28)*	Endocrine, nutrition (52)	Cancer (16)	Cancer (11)

NOTES: Data for the six jurisdictions are from their respective most recent censuses or population surveys: Palau, 1996; Guam and American Samoa, 1990 censuses; CNMI: mid-decade census; FSM: 1994 census; and RMI: 1994 survey. Circulatory: diseases of the circulatory system; cancer: malignant neoplasms; cerebrovascular: cerebrovascular diseases. Except as noted for Guam, accidents include deaths due to accidents, homicides, suicides, and other deaths associated with alcohol and drug abuse.

*Accidents for Guam do not include motor vehicle accidents, which are the fifth leading cause of death.

SOURCE: PIHOA (1997).

Sexually Transmitted Diseases Great concern about sexually transmitted diseases exists throughout the region because the population, particularly the younger population, is considered quite sexually active, many people often have multiple partners, and people do not take appropriate precautions against the transmission of sexually transmitted diseases. By mid-1995, Guam reported 70 cases of acquired immune deficiency syndrome and human immunodeficiency virus infection, the third highest rate in the entire Pacific region. Rates in other U.S.-Associated jurisdictions are lower, but data collection and surveillance are not considered to be completely reliable (Sarda and Harrison, 1995).

Leprosy Leprosy, or Hansen's disease, continues to be a major public health problem in FSM. The World Health Organization (WHO) considers leprosy to be epidemic in FSM. In 1996 WHO and the FSM Department of Health launched a massive program of leprosy screening and treatment in FSM, financed through a donation from a private Japanese foundation (FSM, 1996).

Lytico-Bodig Disproportionate numbers of people in Guam are affected by disease syndromes known as Lytico and Bodig. These neurodegenerative syndromes appear to affect only people born before World War II (Sacks, 1997). On site visits, we were told there are currently 220 confirmed cases. Although the number has been dropping steadily as people with the disease die, the age of onset appears to be increasing. The National Institutes of Health has supported research of these diseases since 1952 and the latest project began in 1997 as a consortium between the University of California at San Diego and University of Guam.

Thyroid Disease In RMI, high rates of thyroid abnormalities are seen as an effect of exposure to radiation from nuclear weapons testing. These abnormalities seem to be slightly greater in exposed populations, and their descendants, than for other populations. Incidences of thyroid cancer also seem to be at higher rates for those closer to testing areas (e.g., Takahashi, Trott, Fujimori, et al., 1997; Palafox, Johnson, Katz, et al., in press).

Disease Prevention and Health Promotion

Immunization Rates As indicated in Figure 2.6, immunization rates for most jurisdictions come close to or actually surpass those for the United States. This results from concerted efforts to improve immunization rates and the availability of funds specially dedicated to immunization.

Nutrition As noted in the earlier discussion on diabetes, poor nutrition has resulted in several major health problems throughout the region. Vitamin A deficiency, a preventable disease that can lead to night blindness, is widespread in FSM, particularly among young children. In fact, FSM has one of the highest rates of vitamin A deficiency in the world (Pryor et al., 1994). Malnutrition is considered the leading cause of death of Marshallese children; in 1989 it accounted for 17 percent of deaths in children under five years of age (Republic of the Marshall Islands, 1988). Obesity is of great concern to all jurisdictions, especially because it is a key determinant of many other noncommunicable diseases and health disorders such as diabetes, coronary heart disease, and strokes. WHO recently reported that the highest rates of obesity in the world were found in the Pacific among Melanesians, Micronesians, and Polynesians (WHO, 1997).

Tobacco Use Of great concern throughout the jurisdictions is widespread tobacco use, and its contribution toward prevalent chronic diseases, such as heart disease and cancer. As the region began to modernize, demand for cigarettes, and other imported items, grew. Today, smoking is more prevalent in the entire Pacific region than in developed countries, and even more common than in many third-world countries. For example, surveys conducted in the 1980s found that one-half of males in American Samoa, and 53 percent of males in Weno, Chuuk were smokers, compared to a little less than one-third of men in developed countries like the United States and Australia. Higher smoking rates were also found in the more urbanized areas. Although cigarette smoking is more prevalent in the Pacific than in developed nations, the surveys also found that Pacific men smoke fewer cigarettes per day than in industrialized countries. Cigarette smoking among Pacific women is much less prevalent than for men (Marshall, 1991). In the CNMI, it is estimated that 18 percent of the total cost of hospital days in 1994 for Chamorro and Carolinian patients was attributable to smoking. This does not take into account the additional costs for outpatient visits, medications, or off-island referral (e.g., anyone with lung cancer was sent off-island for treatment) (Bruss, 1995).

Alcohol and Substance Abuse Average consumption of alcohol in FSM is an astounding two six-packs or 12 drinks per drinking day (i.e., those days on which a person drinks), according to a recent survey (Micronesian Seminar, 1997). Binge drinking is a common practice throughout the freely associated states, most notably on days when government workers receive their paychecks (Marshall, 1979). So, while alcohol may not necessarily be drunk every day, on those days that it is, it is drunk to excess. The total amount of alcohol consumed yearly in FSM (with a total population of 105,506) is the equivalent of almost 1 million *cases* of beer. The survey estimated that FSM has 11,000 problem drinkers, the overwhelming majority of whom were male, and most between the ages of 30 and 44.

CNMI and Guam are both experiencing a rise in drug abuse and related violent outbreaks. The use of "ice" (also known as methamphetamine, a stimulant) is rising, and is often used in combination with alcohol and other drugs. In 1993, arrests for marijuana and methamphetamine were responsible for 22.2 percent of all drug arrests (GHPDA, 1996). On site visits we were told some methampetamine use has already spread to Palau and there are concerns the country may suffer a similar ice epidemic. Hard drug use (excluding marijuana) does not appear to be a problem in the freely associated states at this time; a recent survey found only one current user of hard drugs in FSM (Micronesian Seminar, 1997).

Suicide Suicide has become one of the leading causes of death in many of the jurisdictions. In FSM, suicide has been a significant problem since the early 1970s. In recent years the annual rate was 30 per 100,000 people, and these were mostly young men. Alcohol use was involved with 45 percent of these deaths (Micronesian Seminar, 1997). The suicide rate in Guam, although lower than in FSM, is still higher than the United States. In 1994, the United States had a suicide rate of 11.6 per 100,000 (USDHHS, 1996).

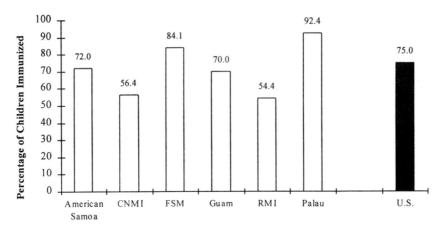

FIGURE 2.6 Percentage of children younger than 2 years of age who are fully immunized, U.S.-Associated Pacific Basin jurisdictions, 1996. SOURCES: Rates for the six jurisdictions are from PIHOA (1997); the rate for the United States was obtained from USDHHS (1996) and is for children 19–35 months who have received the completed set of immunizations; the goal for immunization rates set in *Healthy People 2000* (USDHHS, 1996) is 90 percent. NOTE: "Fully, and completed, immunizations" are defined according to WHO and the Advisory Committee on Immunization Practices.

REGIONAL HEALTH CARE SERVICES OVERVIEW

The health care delivery systems in the Pacific Basin reflect the challenges and strengths unique to the region and the undeniable influence of the nations that have occupied the islands since the turn of the century. In the delivery of health care services numerous challenges must be overcome. These include administrative structures that emphasize acute hospital-based care, the long distances that must be covered to provide care to people in remote areas, dependence on foreign aid, inadequate fiscal and personnel management systems, poorly maintained and equipped health care facilities, the enormous costs involved with sending patients for off-island tertiary care, and shortages of adequately trained health care personnel.

Structure of the Health Care Delivery System

All jurisdictions have planning documents espousing the importance of preventive and primary care, but a review of actual service delivery and budget decisions shows a heavy bias toward hospital-based acute care. This misallocation is often shown as an inverted pyramid in which a small base represents the funding for public health and prevention services, an equally narrow middle represents funding for primary care, and toppling over the entire health care delivery structure, the tremendous uppermost section represents funding for hospital-based acute care and off-island referrals (World Bank, 1994).

With the notable exception of Guam with its relatively well-established network of private clinics, almost all health care practitioners in the region work out of each jurisdiction's central hospital. In the early 1970s, a move was made to decentralize the delivery of health care services. Ambitious training programs were established to train mid-level practitioners, or medexes, who were to go to the outer islands where they in turn would train and supervise the even more peripherally located health assistants in newly opened dispensaries. Unfortunately, this decentralization movement was short-lived. Most of the specially trained medexes were brought back into the district hospitals to fill in for doctors who were in short supply. Similar movements to decentralize the health care delivery system have been tried in the ensuing decades, but have been thwarted repeatedly primarily because of staffing shortages and budget cutbacks. It is hoped with the recent addition of 70 PBMOTP graduates within the freely associated states and American Samoa who have been specifically trained in community health that future efforts at decentralization will meet with greater success.

The Institute of Medicine (IOM) defines *primary care* as the provision of integrated, accessible health care services by clinicians who are accountable for addressing a large majority of personal health care needs, developing a sustained partnership with patients, and practicing in the context of family and community (IOM, 1996). The systems of health care in a few places in the region begin to approach this ideal. Developments in some communities are encouraging, such as the primary care team described in the assessment of RMI in Appendix D.

Most public health care programs rely almost exclusively on U.S. federal funds. Most of that funding, in turn, is rather categorical and may not necessarily focus on the most pressing needs of the individual jurisdictions. The committee is gravely concerned, however, that these programs would simply cease to exist should the United States discontinue its funding.

Administration

Money aimed at improving health has not always been well spent; instances of mismanagement and poor oversight of government funds have been documented in most of the jurisdictions. One frequently mentioned problem on the committee's site visits was the "black hole" of the general government fund. All the jurisdictions pool revenue from all sources into these general funds. Health care expenses—along with all the government's other expenses—are paid out of these huge funds. The directors of health in each jurisdiction have little to no control over how the money in these funds is spent.

It was also reported that it is burdensome for the jurisdictions to apply for, carry out, and report on multiple categorical funding programs. The amount of paperwork required to receive funding from the myriad U.S. grant programs, including those sponsored by the Health Resources and Services Administration (HRSA), creates problems for many of the island jurisdictions, which have limited staff for such tasks. One person may be in charge of applications and reporting requirements for more than 20 different grant programs. Additionally, in several jurisdictions the level of interaction between the individuals who apply for a grant or develop performance indicators and those who actually provide services appeared to be minimal. The purposes of collecting data and setting goals, it seemed, were only to receive funding; they were not seen as methods of improving health care services.

Many people interviewed during the site visits expressed the desire for greater flexibility in the way in which grant funds are administered and used so that federal programs can be adapted to better address local needs. At the same time, they also desired being held more accountable for how money was spent and expressed the need for more technical assistance in applying for and reporting on federal grants.

Coordination

Efforts to achieve better coordination and accountability have been initiated. The Pacific Island Health Officers Association (PIHOA) has been a forum for jurisdictional collaboration. Its membership includes the primary public health officer (who is typically a political appointee) from each of the six jurisdictions. It is administratively headquartered at the University of Hawaii in Honolulu. The aim of this group is to "promote functional and cost-effective solutions to common health service problems, and to collectively achieve improved health status for all island residents" (PIHOA, 1996, p. 1). In addition, PIHOA officially represents the collective health interests of the region; it focuses its efforts on objectives and programs that benefit all of its member states. For example, with financial support from the Centers for Disease Control and Prevention (CDC), PIHOA recently recruited and hired a physician-level regional epidemiologist who will complete an epidemiological needs assessment and develop working surveillance systems and a reporting structure for the entire region.

There is also nascent, but still minimal, coordination among and within different agencies within the U.S. Department of Health and Human Services (HHS). For example, in 1995, the Office of Pacific Health and Human Services was established to improve coordination among HHS agencies including the Office of Public Health and Science, HRSA, CDC, and the Substance Abuse and Mental Health Services Administration (SAMHSA). Similar coordination efforts have been initiated within individual HHS agencies; HRSA has set up an Intra-Agency Workgroup on the Pacific Basin, for example.

Coordination among other federal agencies that serve the region is rather uncommon. This may change as the Compacts of Free Association are renegotiated (e.g., individuals from the Departments of Interior, Energy, Health and Human Services, and State have begun to meet as part of a task force to prepare for Compact renegotiation).

Data Management

Health care data for the Pacific Basin population come from a variety of sources. These include: vital statistics (birth and death certificates), inpatient information (hospital discharge data), outpatient clinic data, public health clinic information (with special information regarding rates of, for example, Hansen's disease, sexually transmitted diseases, tuberculosis, immunizations, and prenatal care), dispensary data, incidences of notifiable (reportable) diseases, and special data sources (e.g., dental clinics, disabled children's services, and laboratories). A number of organizations require and make use of this data including each jurisdiction's own health departments and local government, several U.S. agencies, WHO, and the South Pacific Commission (SPC).

Several problems surrounding these health status data exist, however. They include (1) difficulties in collecting data, especially when births and deaths occur outside the hospital; (2) data overload from too many forms and the collection of more data than can or are being used, (3) lack of data consistency when outside groups or funders request new or slightly altered information; and (4) problems with computerization and automation because of insufficiently trained data personnel, lack of appropriate hardware and software, and unreliable power supplies (O'Leary, 1995).

Health Care Facilities

The state of the health care facilities throughout the region varies considerably. Relatively new and well-maintained hospitals can be found in Saipan, Guam, and Palau. The hospitals in FSM, RMI, and American Samoa are older and are generally in need of much greater repair or rebuilding. Although the majority of health care is provided through the main hospitals, each jurisdiction has a system of primary care sites, often referred to as *dispensaries*, that provide health care to residents in more remote areas and outer islands. These sites often are not well maintained, are staffed by inadequately trained individuals, and usually provide only very limited services. Planning and budgeting for maintenance for all health care facilities and equipment are lacking throughout the region.

Residents are often reluctant to go to these health care facilities because they believe they will often have to wait for long periods before being seen, the equipment and medicine they need may not be available, and the health care professionals may not diagnose or treat the ailments correctly (Flear, 1997b).

Use of Military Health Facilities Although the Compacts direct the secretary of defense to make military facilities available to citizens of the freely associated states, those facilities do not normally treat the patients, except in emergencies. The Naval Hospital on Guam, for example, coordinates with the Guam government in emergency situations such as car accidents when the Naval Hospital is the closest hospital or under emergency conditions that threaten to overload Guam Memorial Hospital emergency rooms, such as the recent crash of a South Korean passenger jet. Tripler Army Medical Center in Honolulu has also been used for medical referrals from the region and is leading many of the region's telemedicine efforts. These facilities can continue to play a key role in the region's health care.

Community Involvement with Health Care Delivery

The community and community involvement have formed the bedrock on which Pacific societies have survived and flourished for centuries. In general, however, over the past 50 years communities in the U.S.-Associated Pacific islands have not been actively involved in the formal provision of health care (which has been primarily through hospitals funded by the United States. in each jurisdiction's main population center). Additionally, the link between health and personal responsibility has not been well cultivated. On the site visits, many islanders noted that the expectation that the United States would take care of health problems, particularly by sending people off-island for diagnosis and treatment, continues to be considered by many islanders as an entitlement. Indeed, many islanders—particularly in the flag territories—have lived or have relatives who live in the United States and, having experienced the health care there, feel they should receive similar services at home.

Traditional health beliefs and practices are thought to have existed throughout the islands for centuries and even today continue to be used either independently from or as a complement to the formal health care delivery systems (Flear, 1997). While many islanders have come to expect and even demand western-style medicine, with its biomedical model and theories about how germs and genetic information are transmitted, many others continue to view illness as being "caused by supernatural transgressions, ancestral transgressions, imbalance of Yin and Yang, or cold or hot forces in the body and other causes not included in the repertoire of western medicine" (Lin-Fu, 1994, p. 296). In some instances western-style medicine is tried only if traditional methods fail. On the site visits, committee members were told that traditional practitioners such as *fao faos* in American Samoa and *surhanos* in Guam continue to provide important health care services to many islanders. Presently, it appears that traditional and western-style practitioners rarely interact or consult one another regarding a patient's treatment (for more details on traditional health practices and beliefs please refer to the assessments in Appendix D).

Nevertheless, some jurisdictions have engendered a good deal of community involvement with health care delivery. For example, volunteers in CNMI have raised more than $750,000 over the past 10 years to buy needed equipment for the Commonwealth Health Center and to support a variety of community health education programs. Guam has several health-related nonprofit organizations such as the American Cancer Society. In Pohnpei, a group of community volunteers banded together and within 2 weeks renovated a building being used as a dispensary (Flear, 1997a). In American Samoa, local businessmen helped to support a fund that bought medicated cream to treat children with scabies. The Red Cross is active in several jurisdictions including Chuuk and American Samoa.

The committee draws particular attention to the role of women and women's groups in achieving improved health. Examples of past contributions of women in the region show promise for their continued involvement and for improved results in the future. Two examples are noteworthy. The Youth to Youth in Health program in RMI, which empowers adolescents to improve their health, was founded, implemented, and managed by female leadership and staff (Youth to Youth in Health, 1996). Women also volunteered critical language interpretation and organizational services at a successful dispensary in Pohnpei (Ruze, 1997).

Similarly, church leaders and heads of community organizations have the potential to provide prevention and treatment services. For example, SAMHSA is now setting up a training program for government employees who will theoretically train the members of community organizations to reach people with substance abuse problems in the villages. This strategy is particularly well-adapted to the situations in FSM and RMI.

Off-Island Care

A legitimate need for medical care for tertiary needs and specialized services not available on an island will always exist. This means travel from one island to another within the same jurisdiction and travel to another country. Such referrals consume much of the total health care budget but benefit only a small number of individuals, often for marginal gains. This is especially notable when terminally ill patients are referred off-island. Off-island care decreases the funding available for care of the general population and for efforts to enhance on-island service capabilities. Off-island referrals also foster a sense of dependency on outside agencies and tend to undermine the public's confidence in on-island providers and facilities. Reportedly, many referrals for off-island care have been given on the basis of political or familial favors rather than clinical necessity.

The problems associated with these referrals are also universally recognized by each of the jurisdictions. Some efforts at cost-saving have been implemented, such as sending people to the Philippines and other countries for treatment, which is less expensive than treatment in the United States. (More detailed information on each jurisdiction's off-island care program is provided in Appendix D.)

Financial Resources

With the exception of Guam, health care services in the region are almost exclusively publicly financed. The U.S. government, in turn, provides the majority of funding for that financing, either directly, as in the case of American Samoa, or indirectly, through Compact and Covenant funding, as in the case of the freely associated states and CMNI. Private health insurance provides most of Guam's health care finances, although it should be noted that government employees are a large part of the market and provide a major source of revenue for the private sector.

Today several federal agencies provide health-related services and funding to the region. The two most important health-related funders are DHHS and the U.S. Department of the Interior (DOI). For example, as indicated in Table 2.3, in fiscal year 1996 DHHS agencies provided approximately $70 million to the region (HRSA, 1996). Of that total amount, HRSA spent $7,314,840 through five different operating divisions (see Table 2.4). DOI provided funds to improve the jurisdiction's infrastructure, including some of its health care facilities, and technical assistance on health-related matters.

Other U.S. agencies providing health-related funding include the Departments of Energy (DOE), Agriculture, Commerce, Education, and Defense. DOE conducts research on and provides financing for the health care needs of those individuals who were exposed to radiation during nuclear weapons testing in RMI. The Department of Agriculture provides child nutrition programs and food stamps to needy citizens and families in the flag territories and administers several agricultural and forestry programs aimed at improving agriculture throughout the region. The Department of Commerce is responsible for conducting the census in the region and has prepared special reports estimating the impact of the Compacts of Free Association in Guam and CNMI. Commerce also operates the National Oceanic and Atmospheric Administration, which has been involved in a number of water and shoreline conservation projects. The Department of Education works in the flag territories to improve basic and secondary education, including for children with special needs. The Department also provides U.S. federal scholarship and grant funds for students throughout the region pursuing higher education. The Department of Defense has military installations with medical facilities in Guam and RMI. Tripler Army Medical Center in Hawaii also sends some medical consultation teams to the region and accepts some patients for specialized tertiary care when they are referred from the jurisdictions. Several civilian action teams also provide some health care services in the region.

Workforce

The 3,142-member health care workforce in the U.S.-Associated Pacific Basin comprises several types of health care professionals including physicians, dentists, mid-level practitioners, nurses, and other allied health professionals. Table 2.5 provides a detailed numeric description of the region's current health care workforce. Although these professionals may have the same title as health care professionals in the United States, their skill levels and the roles they perform can be quite different from their U.S. counterparts. The reliance on expatriate physicians, nurses, and dentists is found in every jurisdiction except Palau and Kosrae (for more detail on each jurisdiction's workforce see Appendix D).

Physicians

The physician workforce includes individuals with M.D. degrees (if trained according to U.S. standards), M.B.B.S. degrees (the British equivalent to the M.D.), and M.O. degrees (if they are medical officers [typically trained in Fiji or the PBMOTP]). Almost all physicians with M.D.s or M.B.B.S.s are expatriate workers brought in on contract.[3] Some have advanced or specialized skills such as in anthesesiology or cardiology, others act as general internists. The ranks of physicians have grown significantly in recent years as a direct result of the training of 70 new indigenous medical officers through PBMOTP described in the appendix to this chapter. Medical officers from the PBMOTP have received five years of formal medical school and clinical experience in Pohnpei and have undergone a two-year internship in their own jurisdiction. They were trained to act as independent practitioners. In their formal training, they were taught the skills to do some types of surgery such as the surgical repair of coral cuts and fish bites, but they were not trained to do more major surgery such as cesarean sections, open chest surgery, or tumor removal.

---

[3]In late 1997, two National Health Service Corps volunteers were working in the region: one in Palau and one in Yap.

TABLE 2.3 Cost Summary for DHHS Programs and Activities in the U.S.-Associated Pacific Basin Jurisdictions, by Agency, FY 1996 (in thousands)

DHHS Agency	Flag Territories			Freely Associated States			Multijurisdiction	Total
	American Samoa	CNMI	Guam	FSM	RMI	Palau		
Administration on Aging	$1,219	$516	$1,987	$0	$0	$277	$0	$4,000
Administration for Children and Families	4,014	2,920	13,925	3,034	1,932	1,762	0	27,587
Centers for Disease Control and Prevention	762	1,336	2,332	1,276	717	805	0	7,228
Health Care Financing Administration[a]	3,181	1,896	15,193	NA	NA	NA	0	20,271
Health Resources and Services Administration[b]	610	629	1,227	781	575	845	2,648	7,315
National Institutes of Health	0	0	600	0	0	0	0	600
Office of the Assistant Secretary for Health	143	72	168	141	67	44	110	745
Substance Abuse and Mental Health Services Administration	431	451	888	566	246	116	0	2,699
TOTAL	$10,360	$7,820	$36,320	$5,798	$3,537	$3,849	$2,758	$70,445

[a]Health Care Financing Costs are Medicaid and Medicare. These programs are not available to the Freely Associated States, and are therefore marked "NA."

[b]Multijurisdictional funds for HRSA are from the Bureau of Health Professions. These funds go directly to the University of Hawaii. In FY 1996, $1.2 million funded the PBMOTP and the remainder funded financial assistance, loans, scholarships, and traineeships for students and health care professionals.

SOURCE: USDHHS (1997).

TABLE 2.4 Health Resources and Services Administration Budget for Activities in the U.S.-Associated Pacific Basin Jurisdictions, FY 1996

Bureau	Total Funding	Flag Territories			Freely Associated States		
		America Samoa	CNMI	Guam	FSM	RMI	Palau
Bureau of Health Professions*	$2,647,805	NA	NA	NA	NA	NA	NA
Bureau of Health Resources Development							
Ryan White Title II Grant	5,790	0	0	5,790	0	0	0
Bureau of Primary Health Care							
Community Health Center Program	975,944	0	0	183,470	142,362	238,479	411,633
Maternal and Child Health Bureau							
MCH Block Grants	2,708,813	510,027	481,693	787,710	538,363	238,011	153,009
EMS for Children Block Grants	47,986	0	47,986	0	0	0	0
Special Health Services Demonstrations	748,128	100,000	99,764	250,000	100,000	98,364	100,000
Office of Rural Health Policy							
Rural Health Outreach Grants	180,374	0	0	0	0	0	180,374
TOTAL	$7,314,840	$610,027	$629,443	$1,226,970	$780,725	$574,854	$845,016

NOTES: MCH = Maternal and Child Health; EMS = Emergency Medical Services.
*Bureau of Health Professions funds go directly to the University of Hawaii. In FY 1996, $1.2 million funded the PBMOTP, and the remainder funded financial assistance, loans, scholarships, and traineeships for students and health care professionals.
SOURCE: HRSA (1996).

Dental Professionals

Of particular concern is the dearth of dentists and other dental health practitioners, particularly in the freely associated states. While 70 dentists are in practice in Guam, only 35 other dentists or dental officers serve all the other jurisdictions. The overall U.S. ratio of dentists to population stands at approximately one dentist to 1,785 persons. In Guam, that ratio is one dentist per 2,218 persons. The rest of the region's ratio ranges from one dentist per 4,306 people in Palau to one dentist per 14,811 people in RMI. Many of the current dental practitioners are expatriates, including one National Health Service Corps dentist in RMI, or are nearing retirement (PIHOA, 1997). Shortages of dentists, dental officers, dental therapists, dental nurses, and dental aides and technicians are already being experienced in all the freely associated states.

Mid-Level Practitioners

The Pacific Basin workforce makes use of several types of mid-level practitioners unique to the region. Individuals licensed as physician's assistants and advanced practice nurse practitioners do work in the region, but only in Guam and CNMI. In the early 1970s, several groups of Micronesian health professionals began to be trained as "medexes," the rough equivalent of a physician's assistant in the United States. Several of these medexes are still practicing today. These individuals typically came with nursing backgrounds and received 2 additional years of specialized training (4 years if they did not have a nursing background). Medexes are trained to be independent community-level practitioners in areas where they are physically separated from higher-skilled physicians. They cover the spectrum of health care services, including disease prevention and health promotion and are able to train and supervise health assistants.

Nurses

Nurses play an important role in the region's health care delivery system. Yet in nearly every jurisdiction, officials reported having a nursing shortage. While most nurses practicing in the flag territories are licensed and practice at the U.S.-equivalent of a registered nurse (R.N.), most nurses in the freely associated states are graduate nurses, meaning they have completed a 2-year, college-level nurse training program. Most nurses in the region practice in the hospital setting.

Many potential explanations exist to explain the nursing shortages. Some nurses move to other jurisdictions in search of better pay and benefits. Because most nurses are women and most physicians and administrators are men, some nurses have quit or relocated when traditional gender roles and expectations have collided with their professional aspirations and training (for example, traditionally most men have come to expect unquestioning obedience from women. This is especially problematic when even the most highly trained and competent female nurse questions a male doctor's decision.)

Health Assistants

Health assistants are individuals trained to staff the outlying dispensaries and non-acute care clinics. They have been trained to respond to most basic first-aid needs, dispense "over-the-counter" medications, and be the link to the central hospital when more serious situations arise. Most are connected to the hospital via shortwave radio. The need for continuing education of health assistants was noted repeatedly on the committee's site visits.

Health Care Workforce Education and Training

Currently, primary and secondary education throughout most of the region does not adequately provide students the skills that they need to participate in the health care workforce. For example, the reading levels of high school graduates in FSM hover around the fourth- and fifth-grade levels and math and science scores are equally poor (Hezel et al., 1997). Almost all of the students entering the Pacific Basin Medical Officers Training Program (PBMOTP) had to take special remedial classes in English, math, and science to come up to speed in these areas. In American Samoa, similar remediation efforts are provided to nursing students. The John A. Burns School of Medicine at the University of Hawaii has established the *Imi Ho'ola* program with the specific aim to improve the academic performance of promising students from the U.S.-Associated Pacific Basin jurisdictions. The program provides remedial coursework to pre-medical students and special support once these students enter medical school. Students from Guam and CNMI—jurisdictions considered to have some of the best educational systems in the region—have participated in the *Imi Ho'ola* program. Some special programs to improve the performance of students who want to go on for secondary education do exist. For example, the "2 + 2" program in CNMI targets high school students interested in pursuing careers in education and links them with mentors and special courses.

TABLE 2.5 Health Care Workforce in the U.S.-Associated Pacific Basin

Position	American Samoa	CNMI	Federated States of Micronesia					Guam	RMI	Palau	Total
			Chuuk	Kosrae	Pohnpei	Yap	Total				
Physicians											
MO	9	1	19	7	17	11	49	0	6	16	81
MBBS	0	0	0	1	2	5	11	0	1	3	15
MD	4	26	4	1	7	1	10	311	19	12	382
Ophthalmologist	0	1	0	0	1	0	1	10	1	0	13
Dental Professionals											
DDS	8	2	5	1	4	0	10	70	4	3	97
DO	2	1	1		2	1	4	0	0	1	8
Dental therapist	1		1	1	0	0	2	0	0	0	3
Dental nurse	0	7	3	3	10	10	26	0	4	6	43
Dental hygienist	0	1	0	0	0	0	0	21	0	0	22
Dental aides	0	14	11	1	9	0	21	221	8	0	264
Dental lab. tech.	0	0	0	0	0	0	0	0	0	2	2
Mid-level practitioners											
Medex	0	0	5	0	4	0	9	0	9	0	18
Physician assistant	0	11	0	0	0	0	0	23	1	2	37

Nurses											
RN	31	139	55	1	63	1	120	648	1	10	949
GN	23	0	0	34	0	12	46	0	79	53	201
LPN	97	9	99	3	49	12	163	164	0	50	483
Aides	12	22	0	0	0	3	3	0	44	0	81
Health assistants	0	0	88	3	15	3	105	0	56	28	189
Allied health workers											
Lab. tech.	20	14	3	6	2	7	18	13	16	7	88
Pharm. tech.	9	17	2	1	2	1	6	57	6	8	103
Rad. tech.	9	14	2	1	2	30	8		5	6	42
TOTAL	225	279	294	64	184	70	612	1,538	260	207	3,121

NOTES: The CNMI data were provided to PIHOA by the CNMI Department of Public Health, but are different from data provided to the IOM committee earlier by the CNMI Department of Public Health. MO = Medical Officer, MBBS = Bachelor of Medicine and Surgery (British equivalent of U.S. MD), MD = Doctor of Medicine, RN = Registered Nurse, GN = Graduate Nurse, LPN = Licensed Practical Nurse, DDS = Doctor of Dental Science; DO = Dental Officer, Dental lab tech. = dental laboratory technician, Lab. tech. = laboratory technologist or technician, Pharm. tech. = pharmacist or pharmacy technician, Rad. tech. = radiology technologist or technician.

SOURCE: PIHOA (1997).

If students do make the grade academically, the financial inability to pay for medical and health education at secondary and graduate institutions of higher education has been a stumbling block to many Pacific Island students. Some federal financial aid programs are available to students. Additionally, in the past students from former trust territories (i.e., CNMI and the freely associated states) were allowed to receive in-state tuition at the University of Hawaii, but this special allowance has been threatened because of recent state budget cuts in Hawaii. Some private scholarship programs exclude Pacific Islanders. For example, the Association of American Medical Colleges, through its Minority Medical Education Program, offers medical scholarships to minority students from underserved areas. Although the program reaches out to native Alaskan and Hawaiian youth, students in the U.S.-Associated Pacific Basin are not currently eligible.

The Yanuca Island Declaration designated both the Fiji School of Medicine and the University of Papua New Guinea as regional centers for postgraduate medical education (WHO, 1995). The DOI, first in cooperation with the PBMOTP and currently with the Pacific Basin Medical Association, has established formal ties with the Fiji School of Medicine to provide postgraduate education to Micronesian medical officers. Several PBMOTP graduates are training there, and many others hope to do so once they have completed their required 2-year internship in their host countries. The Fiji School of Medicine was chosen because the political and social situations in Papua New Guinea are viewed as unstable.

Although continuing medical education is already required for licensure in some of the jurisdictions, it is not a requirement throughout the region. (See individual assessments in Appendix D for more information on licensing requirements.)

Technology and Equipment

The availability of supplies and medicines is also a problem throughout the Pacific Basin, even for the "better" facilities. Shortages of critical supplies were noted in American Samoa, Chuuk, Guam, Pohnpei, RMI, and Palau. Because of budget constraints and other factors, supplies are not routinely or consistently ordered in advance—a process that sometimes requires up to 11 signatures! Generally, the order amount can be altered (typically, lowered) throughout this chain of approval. Thus, with a 3-month lead time from order to actual delivery, an order for X-ray film that started out as 30 cases may actually arrive 90 days later as 6 cases. When the need for such supplies arises, they must then be filled through expensive air freight emergency orders or by emergency referrals to Guam or Honolulu. Some suppliers simply refuse to work with certain jurisdictions because their accounts are so far in arrears. In the past, these bills were routinely paid by the U.S. government or were simply written off by the suppliers.

The equipment and technological capacity for distance education exists. PEACESAT, based at the University of Hawaii, provides all the Pacific Basin jurisdictions with basic satellite and communication services, including Internet access. It is, however, routinely criticized as being hard to use, out of date, and hard to maintain. Technical support from the University of Hawaii was described as minimal. Some people have had difficulty scheduling necessary time slots to use the system. The University of Guam (UOG), through the College of Nursing and Health Science, provides distance education via PEACESAT to help nurses in Micronesia earn advanced credits towards a BSN degree. UOG is also investigating other satellite technology for this purpose. Northern Marianas College has recently invested in satellite technology of its own and is already offering several of its Saipan-based classes to students on the island of Tinian.

APPENDIX

Pacific Basin Medical Officers Training Program

In the early 1980s many of the physicians in Micronesia (primarily RMI, FSM, and Palau) were nearing retirement. With few students from the region in medical school and even fewer of those who did finish their medical training returning home to practice medicine, a potentially devastating physician shortage seemed imminent (Pretrick, 1997). Officials from the affected jurisdictions worked together with the John A. Burns School of Medicine at the University of Hawaii to create a response. The solution—the formation of a regional medical officers training program based in Pohnpei—received a commitment of 10 years of funding from the U.S. Department of Health and Human Services' Health Resources and Services Administration. The Pacific Basin Medical Officers Training Program (PBMOTP) enrolled its first class of students in 1986.

Overview of PBMOTP

Student Recruitment and Retention

With the exceptions of Guam and CNMI, each jurisdiction sent students to PBMOTP. Recommendations for students to be enrolled in the program were solicited through admission boards established in each country (and all four states of the FSM). The department of health within each jurisdiction administered and oversaw these boards. PBMOTP provided students with room and board, transportation, books, and a small stipend for living expenses (some of which could be sent directly to the students' spouses back home). Students were also given time off twice a year for home visits.

Over the 10-year span of the PBMOTP, 170 students were accepted into the program; 58 percent (99 students) did not complete the training. Several reasons explain this dropout rate. The vast majority (about 80 students) dropped out early for academic reasons, including trouble communicating in English. Problems with alcohol abuse, stealing, and other behavioral problems resulted in some students being asked to leave, particularly in the early years of the program. Separation from families and the resulting depression and loneliness also led some students to drop out. Nevertheless, some of the students who dropped out of the program are currently providing health care in the region as either health assistants (12 individuals) or medexes (10 individuals).

In the later years of the program, several factors brought about increased retention rates. The admissions board nominated better-qualified individuals. The program hired top-quality faculty, many of whom were Pacific Islanders with good credentials and who served as positive role models (as did the medical officers who had already been graduated from the program). The faculty also developed a screening test for English as a second language, which quickly and accurately predicted a student's ability to perform academically using English. Students who did not score well on the screening test were encouraged to reconsider their decision to participate in the program. Others were required to take a survival skills program for the first 6 months of the program that provided remedial courses in English, math, and science.

The program's recruitment and retention of women is particularly noteworthy. Special efforts were made to recruit capable women to the program because they were so underrepresented in the professional workforce. The program did make some special recruitment trips to the jurisdictions but recruitment was a continuing process during PBMOTP Advisory Board meetings and other regional health training activities and conferences. During the clinical workshops for the Micronesia Otitis Media Training Project (funded by HRSA) and the PBMOTP's on-site clinical rotation in Chuuk during 1990–1991, bright women were noted and then actively encouraged to apply to the program. Directors of Health were informally encouraged by PBMOTP faculty to find and recruit able women. Additionally, over the course of the program, six well-respected female physicians and three nurse midwives worked as program faculty. They acted as professional role models and as confidants and advocates to female students, bringing attention (formally and informally) to sexual harassment and abuse within the program.

Female PBMOTP students also had a higher retention rate than that of the men. In the last graduating class, more men than women were initially admitted, but by graduation women outnumbered the men (and women took all the top academic prizes). About half of the female PBMOTP students were married and about two-thirds of all the female students were mothers. In fact, six students had babies while in the program and they all went on to complete the program. While at the PBMOTP, most female students were separated from their families and most were able to handle the separation quite well.

Curriculum

In about 1990 the PBMOTP teaching approach changed from being teacher-centered to being "problem-based." This switch placed more of an emphasis on community health training, rather than hospital training, and moved away from a focus on classroom lectures to student research. About half of the program was academic and the other half consisted of supervised clinical experience. Students were required to conduct self-directed research projects and to staff several "clinics without walls" located in several communities near the school. This change is credited with invigorating the program and seemed to work well at motivating students to become more actively involved with their education: to think critically and to solve problems. This problem-based type of approach is now being implemented in a variety of educational institutions throughout the Pacific, including the Fiji Schools of Medicine and Dentistry and the University of Guam School of Nursing.

Career Ladder Program

PBMOTP also provided students with a graduated career ladder. During years 1 to 3, students engaged in self-directed learning with a focus on using clinical skills in community health settings. The students assisted in public health and community medicine clinics throughout Pohnpei (thereby fulfilling some service needs in Pohnpei State). After the first year, students became health assistants. Upon completion of the third year, students were certified as assistant medical officers (AMO) and licensed as medexes. During years 4 and 5, students focused on inpatient medicine, primary care, mental health, and more projects in the community. During this time, students also returned to their home jurisdiction for a 4-week clinical inpatient internship at their local hospital. Returning home also helped students reconnect with their communities and allowed them to spend time with their families.

Upon successful completion of the 5-year program and a final qualifying examination process, students became medical officers and received a diploma in Community Health, Medicine, and Surgery from the John A. Burns School of Medicine at the University of Hawaii at Manoa. Once they completed a 2-year internship with the respective department of health, the graduates took a registration examination, and if they passed, they became fully licensed medical officers.

Results

PBMOTP fulfilled its mission to provide a locally trained physician workforce and to eradicate the chronic physician shortage of the 1980s. Over the course of the 10-year program, 70 students graduated and are now practicing in the region (see Table 2.6). Half of these graduates are women (Head, 1997). The total cost of the program tallied almost $15,000,000 (see data in Table 2.7). This money had the secondary effect of providing health care for the local communities.

The founders of PBMOTP instituted several policies to address concerns about medical officers leaving the islands once they had completed training. For example, the medical officer license granted to PBMOTP graduates is recognized only in FSM, RMI, and Palau. Students who graduated were guaranteed jobs with their home governments at a salary considered quite good for the region (although not as high as those for expatriate contract workers). To date, all graduates have remained in the region (although some have married classmates and moved to islands other than the jurisdiction that sponsored their training) (see Table 2.8). However, as Compact funding decreases, governments in the region are reducing costs. Whether a medical officer continues to practice is up to his or her government's ability and willingness to support them.

TABLE 2.6 Number of Physician Graduates, by Year of Graduation

Year	Number of Graduates
1992	15
1993	8
1994	14
Early 1996	8
Late 1996	23
1997	2
TOTAL	70

SOURCE: Dever (1997).

TABLE 2.7 PBMOTP Budget, FY 1986 to FY 1996

Year	Budget
1986	$894,358
1987	716,288
1988	884,954
1989	1,165,000
1990	1,660,000
1991	1,577,000
1992	1,800,000
1993	1,700,000
1994	1,500,000
1995	1,482,000
1996	1,456,000
TOTAL	$14,836,426

SOURCE: Dever (1997).

TABLE 2.8 Number of PBMOTP Graduates Currently Working in U.S.-Associated Pacific Basin Jurisdictions

Jurisdiction	Total
American Samoa	9
FSM	42
Chuuk	16
Kosrae	7
Pohnpei	13
Yap	6
RMI	6
Palau	13
TOTAL	70

NOTE: Some graduates were sponsored by one jurisdiction but are currently working in another jurisdiction. So, for example, American Samoa only sponsored seven students at the PBMOTP, but two students from other jurisdictions moved to American Samoa to practice, resulting in nine total PBMOTP graduates working there.
SOURCE: Dever (1997).

Many of the female PBMOTP graduates will likely assume leadership positions in the jurisdictions' health care delivery systems. This is widely viewed as a positive step. As stated earlier, women are underrepresented in the professional physician workforce. Their absence was particularly problematic given the many cultural restraints and taboos associated with indigenous male health professionals caring for female patients, especially regarding reproductive health issues.

Continuing Education and Training for PBMOTP Medical Officers

Now that the medical officers have received their basic training and have begun to practice formally, they will be the mainstay of the physician workforce, particularly in Palau and throughout the FSM states. Nonetheless, they will still need continued education and training. Such activities help medical officers improve their skills, which will help build the public's confidence in their competence and capabilities. The lack of ongoing education and training was given as one of the problems faced by the region's older cohort of medical personnel. PBMOTP graduates will need to share the knowledge they have learned in school and in practice with other health care workers. Their positions as role models will help them achieve this goal, but such community outreach efforts will also require continuing education and training.

Replication of PBMOTP

Although the committee believes that PBMOTP was a remarkable success, it does not see any current application for replication of such a program within the mainland United States. It does believe, however, that the model would serve other developing nations well as they seek to train indigenous people to be health care practitioners. The U.S. Department of Health and Human Services and the U.S. Agency for International Development are encouraged to share information about the program's design and curriculum with officials in those developing countries. The following aspects of PBMOTP are especially worthy of duplication:

- having community leaders nominate students,
- using a problem-based teaching approach,
- emphasizing the development of clinical skills from the start of the program, and
- using an approach that allows student to climb a career ladder as they increase their skill levels.

However, if such a program were to be replicated, the committee recommends working more closely with existing local educational institutions rather than setting up completely new—but temporary—institutions.

3

Charting a Course for the 21st Century: A Strategic Plan for Future Health Initiatives in the U.S.-Associated Pacific Basin

The committee strongly recommends continued U.S. involvement and investment in the region's health care. The nature and scope of this involvement and investment, however, must change. Beyond merely providing health care—a great challenge in its own right—the United States and the island communities must work together with a renewed sense of partnership to produce improved health of Pacific Islanders.

To achieve this goal of improved health the committee recommends a four-pronged approach:

1. Adopt a viable system of community-based primary care and preventive services.
2. Improve coordination within and between the jurisdictions and the United States.
3. Increase community involvement and investment in health care.
4. Promote the education and training of the health care workforce.

The committee believes that priority should be given to the first two goals—adopting a viable system of community-based primary care and preventive services and improving coordination within and between the jurisdictions and the United States. To a certain extent the other two goals of increasing community involvement and strengthening the health care workforce flow naturally from these first two priorities. The committee cautions, however, that if the two main priorities are not given serious attention by all parties involved, the desired goal of improving the health care systems in the region and ultimately the health of the island populations will not be achieved—even if the full set of other recommendations are fully implemented.

Implementation of some of the approaches recommended below has already begun in the jurisdictions. At times these new approaches have met with success, other times they have failed. The committee underscores the importance of working on several approaches at the same time. Taken individually the approaches will not have the same impact as the collective efforts and opportunities for potential synergy will be lost.

Although it was beyond the charge, expertise, and capability of the committee to make detailed estimations of the costs of implementing these recommendations, the committee believes most of the costs can be covered through the reallocation of current levels of health care funding—especially as a more locally sustainable and viable system of community-based primary care and preventive services is adopted. In the 1993 landmark report, *Investing in Health*, the World Bank calculated the cost of providing a minimum package of public health and essential clinical services in low- and middle-income countries (World Bank, 1993). All the U.S.-Associated Pacific Basin jurisdictions are considered middle-income countries. In 1990, the cost of such a package for middle-income countries was $22 per capita. All the jurisdictions currently have considerably higher health budgets per capita as shown in Table 3.1 (PIHOA, 1997).

TABLE 3.1 Total Health Budget Per Capita, U.S.-Associated Pacific Basin Jurisdictions

Jurisdiction	Total Health Budget Per Capita
American Samoa	$369
CNMI	614
FSM	132
Chuuk	92
Kosrae	151
Pohnpei	143
Yap	178
Guam	510
RMI	128
Palau	320

SOURCE: (PIHOA, 1997).

ADOPT A VIABLE SYSTEM OF COMMUNITY-BASED PRIMARY CARE AND PREVENTIVE SERVICES

Fundamental reform of the ways in which health care services are provided is occurring throughout the world. Reasons for this reform movement include the urgent need to decrease overall costs and increase cost-effectiveness, the desire to improve service delivery, and ultimately, the will to achieve better health outcomes. This wave of reform has also hit the U.S.-Associated Pacific Basin. All the jurisdictions are struggling to design and implement health care delivery systems that are more locally sustainable, that provide more consistent and higher-quality services, and that more effectively improve the health of their residents.

Improve the Island's Physical Infrastructure

Basic sanitation, access to potable water, and other essential preconditions to good public health remain of great concern, particularly in the freely associated states. Power outages occur fairly regularly in the Federated States of Micronesia (FSM) and the Republic of the Marshall Islands (RMI). Poorly maintained and overcrowded roads contribute to accidents and injuries. Continued inattention to infrastructure development by the local governments is viewed as a barrier to health and a factor contributing to disease and illness. The committee believes that investment in infrastructure development is critical to the health and well-being of the people in the region. Assurance of potable water supplies, adequate sanitation, and reliable electricity throughout the jurisdictions must be a clear priority for both the aid-giving agencies of the U.S. federal government and the communities that strive to create healthy islands.

Invest in Preventive and Primary Care and Public Health

The committee underscores the vital importance of investment in preventive and primary care and in population-based public health care in the region. This investment does not necessarily require a great deal of new funding. Currently almost all of these activities are funded entirely by the U.S. federal government. The committee believes that it is important to have local governments adequately fund these functions and provide more funds through local sources. What it will require, therefore, is the reallocation of existing funds—a difficult process in any health care system. In some jurisdictions, it will also require the reorganization of delivery systems to better integrate the acute care, primary care, preventive care, and public health sectors.

To combat the inappropriate and discretionary use of funds for health care services, the committee recommends that each jurisdiction place the funds reserved for this purpose in a separate cost center within the overall health budget. This is another difficult change because most jurisdictions' governments are reluctant to give up their control and access to these funding streams. One approach may be to limit the amount of money spent on tertiary care (both for care on-island and for referrals for off-island care). In Papua New Guinea, for example, public spending on hospitals has been successfully limited to 40 percent of the Ministry of Health's recurrent budget (World Bank, 1993).

Maintain or Establish Basic Health Care Standards

U.S. licensing regulations for health care facilities, providers, and services in Guam, the Commonwealth of the Northern Mariana Islands (CNMI), and American Samoa should generally be maintained. These jurisdictions receive U.S. federal funding to provide Medicare and Medicaid services. The quality of these services should be roughly the same—no matter where a person receives them. Adherence to U.S. standards is necessary to ensure consistency of quality and accountability of U.S. taxpayer funds. Modification can be considered when a jurisdiction provides appropriate justification, stating its rationale and the fiscal implications. Such modification should apply only to a particular facility and should be subject to review by the Health Care Financing Administration (HCFA). HCFA staff should provide on-site assistance to those jurisdictions that demonstrate a willingness to make corrections. HCFA oversight of facilities that continue to be in noncompliance should be done on a quarterly basis. Target dates for decertification must be adhered to strictly. If a jurisdiction believes that adherence to U.S. standards, even with modifications, is too costly or culturally inappropriate, then perhaps it should fundamentally reconsider its participation in these U.S. federal health care programs.

Each of the three freely associated states is strongly urged to establish its own standards for available resources and appropriate technology, including provisions for the licensure of health care providers and legislative practice acts. HCFA should be responsive in assisting these jurisdictions with developing such standards. This may require new legislative authority for HCFA.

Develop a Regional Health Information System

Accurate and informative data are critical for health care reform. The lack of good data hampers policymakers' and administrators' ability to analyze the current situation, set priorities, and plan for the future. The committee encourages each of the jurisdictions to participate in the development of a standard regional health information system. This system would be the repository of information needed to analyze the health care system and to make assessments of how best to proceed with reforms. The system should be able to track health outcomes and progress toward *Healthy People 2010* goals that the jurisdiction has revised to more appropriately capture its unique circumstances and disease burdens (U.S. Department of Health and Human Services, 1997). However, the importance and practicality of developing compatible data systems that use the same defined terms, age groupings, and sampling frequencies across the jurisdictions cannot be underscored enough if regional data analysis is to be made possible.

The Pacific Island Health Officers Association (PIHOA) is the logical candidate to provide the leadership for such an undertaking, with appropriate technical assistance provided by various U.S. federal health agencies. Such efforts should be coordinated with similar data improvement efforts and other requirements of the World Health Organization (WHO), South Pacific Commission (SPC), and other users of regional health data.

Reform Health Care Facility Management

Health care facilities throughout the region are generally in poorly maintained buildings, experience chronic and commonplace shortages of vital supplies and equipment, and have outdated and broken equipment. To improve this situation, the committee recommends that each jurisdiction's government provide officials at the hospitals and other health care facilities with greater control over finances. For example, give the hospital administration the freedom to use hospital fees to purchase drugs and other supplies and to reorganize the ordering system and payment schedules to ensure that critically needed supplies are readily accessible on-island. Each of the jurisdictional governments must also be required to establish an annual budget with separate cost centers for health services facility maintenance and repair, equipment and supplies, salaries, and in-service training. This budget should be tied to an item-by-item work plan on an annual basis. Without the annual work plan it is difficult to determine if the budgeted figure is reasonably close to actual performance requirements. Financial assistance from U.S. funding sources for facility and equipment repair should be tied to the preparation and completion of these annual work programs, and to the inclusion of partial financing with local funds.

The chief health administrator in each jurisdiction should receive advanced training through a certificate or degree program in health administration and health systems management. The training should include coursework taken at institutions of higher education that offer high-quality programs in this subject area combined with practical applications and fieldwork within the administrator's own jurisdiction. A "buddy system" model of having the administrators who have received their training go to another jurisdiction in the region to provide on-site technical assistance should also be used. Finally, training and technical assistance will also be needed to assist other administrative and clinical staff transition to these new management approaches and practices.

Promote Prudent Privatization

Many of the jurisdictions have begun or are considering contracting out and privatizing many of their health care services in an effort to reduce costs and increase health care options for residents. Several physicians have entered private practice and some independent pharmacies and clinics have opened in recent years. Although the committee is generally supportive of such efforts, it also cautions that such arrangements should be carefully considered. Previous attempts at privatization in other economic sectors have not always been successful. Reasons for failures include poor or inadequate collection of fees and the low productivity of some personnel. Additionally, in most jurisdictions privatization will require new legislation dealing with issues such as malpractice and the private use of public facilities.

Technical assistance should be provided (through local institutions of higher education and cooperating professional societies) to the prospective private businesses. These businesses should be required to submit proposed business plans that include realistic fee collection goals and carefully considered policies regarding supervision and other personnel matters.

Rethink and Restrict Off-Island Tertiary Care Referrals

Each jurisdiction has begun to address the issue of off-island referral for tertiary care, but many difficult choices and changes still need to be made. The committee believes each jurisdiction must fundamentally rethink its government's financial support of off-island referrals. Clearly, there will always be a legitimate need for such referrals; the issue is who should pay for them. The committee recognizes the important steps that several jurisdictions have taken to reduce the costs of these referrals such as instituting co-payments; requiring greater cost sharing; creating referral committees of health care practitioners who use specific criteria and protocols to determine when referrals for off-island tertiary care should be made; and using competitive bidding. However, the committee also encourages each jurisdictional government to move away rapidly from providing financial support for such referrals altogether and to consider the development of insurance systems, possibly private-market insurance, or other funding mechanisms to cover such catastrophic health care costs.

Planning assistance should be provided to each jurisdiction willing to reduce off-island tertiary care referrals to assist with developing locally acceptable and sustainable methods and timetables for reducing the health care funding being spent on off-island referrals. Ideally, the money formally used on off-island tertiary care can be used for primary health care, wellness and health promotion, health education, and on-island acute care to help reduce the need for such referrals in the first place.

IMPROVE COORDINATION WITHIN AND BETWEEN THE JURISDICTIONS AND THE UNITED STATES

To maximize scarce resources and minimize wasteful duplication of efforts, the committee calls for greater coordination and collaboration as well as improved management on both sides of the Pacific. Such coordination has begun, but it must be more focused and more consistently pursued and supported. Figure 3.1 provides an overview of the committee's recommendations for achieving this goal.

STRATEGIC PLAN FOR FUTURE HEALTH INITIATIVES

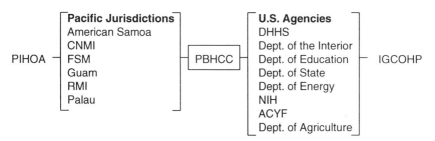

FIGURE 3.1 Overview of recommended organizational arrangements to improve coordination between the U.S.-Associated Pacific Basin jurisdictions and the United States.

ROLE OF THE UNITED STATES

Use of Block Grants That Require Meaningful Measures of Accountability

U.S. federal agencies are encouraged to use block grants or to consolidate grants whenever feasible. The committee particularly encourages the use of common grant applications and consolidated reports to sponsors. The emphasis should be on achieving greater flexibility and efficiency with well-defined measures of accountability and common data systems. One possible approach is to have an agency set broad goals that it would like to see achieved in the region and then allow the individual jurisdictions to decide how they will go about achieving those goals. Outcome measures for assessing success would be developed jointly with the federal funding agency and the jurisdictions. This emphasis on results-oriented management and accountability is increasingly becoming a focus for all government programs. It is also well-suited to the region because in practical terms federal officials often cannot monitor Pacific Basin jurisdictions closely because of the prohibitive travel time and travel costs involved.

An example of such an effort would be to allow all the funds from the Centers for Disease Control and Prevention (CDC) and Health Resources and Services Administration (HRSA) aimed at preventive services (i.e., immunizations, screening, health education, etc.) to be consolidated into one block grant. Jurisdictions would then submit one application and report their progress back to both agencies on an annual basis using agreed-upon outcomes-based indicators and one reporting form.

Consideration of Multiple Uses of Military Facilities

The committee recommends that officials from all U.S. military health facilities in the region enter into dialogue with the jurisdictions to determine the optimal means of sharing regional resources and training opportunities to serve local populations.

Continuation of Funding of Research Projects in the Region

In the past the United States has funded scientific studies in the Pacific, such as studies of Lytico-Bodig on Guam (funded by the National Institutes of Health), and several studies on the health effects of radiation exposure in RMI. These studies have helped scientists make important new discoveries and gain greater understanding of subjects ranging from the biological basis of Alzheimer's disease to the effect of nuclear radiation exposure on the thyroid.

The studies have not been universally viewed as successful in the islands, however, and many islanders are rightfully skeptical and disillusioned with U.S. research. In Guam, for example, epidemiological surveys in 1953 and 1987 caused considerable dissatisfaction among many Guamanians who felt they were not fully informed about the study, not given the opportunity to give proper consent, and had their confidentiality violated (Workman and Quintana, 1996). One of the most contentious issues in RMI relates to the effects of radiation exposure on its people (Simon, 1997). Although in the Compact of Free Association with RMI the United States accepted responsibility for the effects of radiation exposure during the period of U.S. nuclear weapons testing, and accordingly set aside trust funds and other monies as compensation, there is still uncertainty about the degree to which those tests affected the health of those directly exposed and among the people living outside the limited four-atoll region specified in the Compact. Documents that were declassified by the U.S. Department of Energy in 1993 indicated a much higher level of contamination of the islands than was previously recorded. It should be noted that workers who cleaned up test sites came from throughout the region and not just RMI. Additionally, recent reports linking health effects from nuclear weapons testing in Nevada to places as far away as New York have raised concerns from other island jurisdictions about possible radiation exposure (National Cancer Institute, 1997).

Despite these problems and obstacles, the committee believes research in the region needs to continue. Changes must be made, however, in the topics being studied and the way research is conducted. Lytico-Bodig and thyroid cancer affect only a few hundred individuals. Research is needed to examine the major causes of morbidity and mortality in the region such as diabetes, substance abuse, tuberculosis, nutritional deficiencies, women's health issues, and other topics identified by Pacific Islanders as important to improving their overall health. Studies on health systems development that take into account the unique social and cultural belief systems in the U.S.-Associated Pacific Basin jurisdictions are also strongly encouraged. Finally, the committee also encourages greater interagency coordination to support existing research and monitoring such as studies of the health effects from radiation exposure throughout the Pacific Basin jurisdictions, albeit with continued specific attention to the Marshall Islands.

Research should be carried out using accepted ethical methodologies for participatory research, and should be cleared through an institutional research board in the appropriate jurisdiction. Efforts should be undertaken to appropriately involve islanders in the planning of any research. The committee further recommends that research be conducted not only by clinicians, but also by such researchers as public health workers, medical anthropologists, and epidemiologists. Taken together, such research would not only help health care workers, decisionmakers, and policymakers in the Pacific, but also U.S. funders, to better match health programs with the unique needs of the region.

Establishment of an Interagency Governmental Committee on Pacific Health

The committee believes that coordination of U.S. funding for all health-related activities in the Pacific Basin is needed to increase the coherent and consistent application of rules, regulations, and accountability requirements for expenditures, which should be based on the previously discussed outcomes measures. The development of the HRSA intragency work group has shown that improvement is possible. As the Compacts begin to be renegotiated, such collaboration at the federal level will be critical.

In its deliberations, the committee considered several possible organizational arrangements to address this need for improved coordination of federal programs. These deliberations were informed by the IOM's Board on International Health recent report, *America's Vital Interest in Global Health* (IOM, 1997a), which also noted that "fragmentation of governmental responsibilities, divisions of authority between domestic and international health activities, and lack of coordination among U.S. governmental agencies and with the nongovernmental sector impede progress toward global health" (p. 7). Consolidation of all programs and funds into one agency, for example, was determined not be practical given the very different and complex roles each agency performs and the special expertise available from each. However, the board recommended that, because of its unique scientific and health expertise, the U.S. Department of Health and Human Services (DHHS) should act as lead agency to coordinate global health strategy and priority setting across the federal agencies.

The committee therefore recommends the establishment of an Interagency Governmental Committee on Pacific Health (IGCOPH). This committee should be headed by the Secretary of Health and Human Services and should include representation from all federal agencies that fund health-related activities in the region, including DOI, DHHS, and the Departments of Agriculture, Education, Energy, State, and Defense. The goal of the committee will be to ensure coordination of health programs, administrative oversight, and technical assistance to the region. IGCOPH should submit an annual report to the President and to Congress detailing the progress it made in achieving its goals.

IGCOPH Composition

The committee should be chaired by the Secretary of Health and Human Services or his or her designee. The committee believes it is important to have senior representation not only from DHHS, but from DOI, especially given the considerable resources DOI provides that directly affect the health of the populations and health care systems in the region. The committee should also include representation from each of the federal agencies that fund health-related activities in the region, including, but not limited to the following:

- U.S. Department of Health and Human Services (Administration of Children, Youth, and Families; HCFA; HRSA; CDC; and the Substance Abuse and Mental Health Services Administration),
- U.S. Department of the Interior (Office of Insular Affairs),
- U.S. Department of Agriculture,
- U.S. Department of Commerce,
- U.S. Department of Defense,
- U.S. Department of Education,
- U.S. Department of Energy, and
- U.S. Department of State.

IGCOHP Tasks

The specific responsibilities of the committee should include:

- ensure coordination of health programs, administrative oversight, research and technical assistance to the region;
- propose and support the rationale for any future federal health initiative for the Pacific Basin jurisdictions before its implementation and be explicit about the goals and objectives of such initiatives;
- coordinate coherent and consistent rules and regulations on federal health expenditures in the region;
- review agency grant reporting requirements and emphasize consistency on performance measures among agencies;
- identify resources throughout the federal agencies that could provide technical assistance with health sector reform; and
- report annually to the President and to Congress on the committee's progress.

ROLE OF THE ISLAND JURISDICTIONS

Recognizing PIHOA's contribution to regional public health coordination, PIHOA is encouraged to continue its significant public health focus and its mission to promote interregional sharing of resources. PIHOA is further encouraged to (1) develop a regional health information system to promote a shared version with standard nomenclature, (2) review purchasing practices and encourage shared purchasing and volume buying to decrease costs and share resources in emergencies, and (3) identify technical assistance and consulting strategies that promote the prudent use of the expertise available within the region.

INTERFACE BETWEEN THE UNITED STATES AND THE ISLAND JURISDICTIONS

Establishment of a Pacific Basin Health Coordinating Council

Finally, the committee recommends that the governments of the United States and the six island jurisdictions establish or designate a nongovernmental organization in the region to coordinate health affairs and facilitate collaboration between the United States and jurisdiction governments. This Pacific Basin Health Coordinating Council (PBHCC), would meet quarterly and would report annually on the progress of health sector reform in the U.S.-Associated Pacific Basin to the President of the United States, the U.S. Congress, the chief executive officer and legislature of the island jurisdictions, IGCOPH, and PIHOA. PBHCC should have a small permanent staff. The establishment of such a council is not meant to create yet another layer of bureaucracy; rather, it is envisioned as the catalyst for pragmatic health reforms and the watchdog for greater accountability of all parties—in the United States and the region.

PBHCC Composition

The 14-member council should have representation from three different groups: 4 representatives of the U.S. government and IGCOPH, a representative of each of the six island jurisdictions' governments, and a total of 4 private citizens, 2 each from the United States and the island jurisdictions.

PBHCC Tasks

What projects are undertaken by the PBHCC will need to be determined cooperatively with all the parties involved. The committee recognizes the differences in budgets, health care services, personnel, and program directions between the U.S. flag territories (American Samoa, CNMI, and Guam) and the freely associated states (FSM, RMI, and Palau). Therefore, as it undertakes the following tasks, PBHCC should consider grouping jurisdictions accordingly. The PBHCC could:

- help to develop health care priorities in the six jurisdictions that take into account burden of disease and cost-efficiency criteria, as well as priorities developed by the community.
- ensure coordination of U.S. health programs, activities, and funding streams within the U.S. flag territories and the freely associated states.
- review each jurisdiction's progress toward meeting specific health outcomes objectives as required under various U.S. grants.
- assist with the simplification of filing and reporting formats and forms for various U.S. grants.
- facilitate training in health administration for ministers or directors of health and for members of jurisdictional health authorities or boards; and
- identify and establish working relationships with U.S. federal agencies and international organizations and other aid donors (e.g., the World Health Organization, Asian Development Bank, and South Pacific Commission) that could provide technical assistance resources for health care reform.

PBHCC Funding

Funding for this nongovernmental organization must come from a variety of sources. As described in Chapter 1, the U.S. federal government has several vital interests in investing in the region's health and ensuring that the money it provides is spent wisely. In keeping with several of its other recommendations, the committee underscores its belief in the vital importance of having the local jurisdictions provide financial support to this endeavor. The committee also sees a role for private organizations and foundations—both inside and outside the region—to play in funding the PBHCC. All these funding partners must believe that they have a stake in the PBHCC's work and will benefit from the results that work produces.

Several possible funding mechanisms exist. The committee suggests a few options here, but ultimately the various funding partners must collectively determine exactly how each will pay. The U.S. federal government might consider contributing a fixed percentage of all funds it provides to the region. Similarly, each of the island jurisdictions may decide to base its funding on a fixed percentage of its total health care budget or on a fixed percentage of the total funds it receives from the U.S. federal government and international sources.

INCREASE COMMUNITY INVOLVEMENT AND INVESTMENT IN HEALTH CARE

The committee believes that any attempt to improve the health care in the islands of the Pacific Basin must tap into the strengths and resources of the community—if the improvements are to be meaningful and sustained. Fostering an environment that enables households to improve the health of their members, particularly by promoting the rights and status of women, is seen as one of the best ways to improve global health (World Bank, 1993). This focus on women is particularly apt in the Pacific, given the central roles of women and girls in many of the island societies. For example, women are the primary caregivers for both young and old. This is especially important in the jurisdictions with high rates of fertility and infant mortality, where health education about health promotion and disease prevention can lead to vast improvements in health status. Women are also usually responsible for meal preparation and therefore can have a huge influence on the entire population's diets.

The committee acknowledges the differences in the institutional capacities of each of the jurisdictions and in the cultural norms and functioning of individual communities. No one paradigm of community involvement applies to all island cultures equally or necessarily appropriately. Health services must be aligned to each community's needs and congruent with each unique culture, with special attention given to the most vulnerable groups. Each community will have to determine how best to achieve a level of involvement and investment that is needed to truly make a difference in its health.

Island communities should consider taking some fundamental steps, including the following: establish a jurisdictional health authority or board, develop a health improvement benchmarking process, use nongovernmental community organizations to provide health services, and increase community involvement with primary care sites. These steps are described in more detail below.

Establish Jurisdictional Health Authorities

Jurisdictions should create, through local legislation and community input, an independent authority or board to oversee the administration of the health care system, plan and prioritize health initiatives, and provide accountability. Such an authority or board would oversee the budgets of all the health services, agencies, hospitals, primary care sites (dispensaries), and programs under its direction through the development and utilization of sound annual budget practices, monitoring systems, and timely annual audits. Such enforcement and monitoring powers should help to depoliticize health policy decisions.

These authorities or boards should be independent of the government. Funds for all health care programs should be kept separately from the government's general fund. Nominations for membership on these authorities or boards should come from the jurisdictional government and local communities. The health authority or board should include both men and women, and community volunteers such as businesspeople, clergy, educators, and health care professionals. The secretary or director of health would report directly to the health authority or board, although the president or governor of the jurisdiction would retain the right to veto decisions. Regular training in health administration should be provided to health authority members. This should include training to help Health Authority members understand and interpret the methods and findings of health research in the science of epidemiology and public health; conduct focus groups and analyze focus group information; improve their community organizing skills; and communicate information to the public.

Develop Health Improvement Benchmarking Process

Health improvement benchmarking is a method of comprehensive, long-term health planning. It involves determining how well a community is doing in a certain area of health and then deciding what should be done to improve that area to a desired level. One of the committee's charges was to develop assumptions about benchmarks established by earlier studies of the region's health care systems conducted by the University of Hawaii (1984, 1989) for their adequacy in assessing the needs and health status of the populations of the U.S.-Associated Pacific Basin. The committee was then to modify or re-create these benchmarks as necessary to assess the accomplishments, adequacy, and shortcomings of the health services programs and related health interventions in each jurisdiction.

During the site visits committee members were told that health planning and goal setting are often empty exercises done only to satisfy grant requirements. People throughout the region reported that using such activities as a means of identifying, prioritizing, and ultimately improving community health rarely occurs. Therefore, rather than actually going through yet another empty exercise, the committee decided to recommend the development of methods to more actively and meaningfully involve the community in health planning and the implementation of new goals and objectives to achieve improvement.

The committee recommends that individual jurisdictions and communities establish a process for determining which health issues are of greatest concern and how they will monitor their progress in addressing those issues. Health services must be tailored to the unmet needs in the region, especially those of women and vulnerable populations such as children and those with substance abuse or mental health problems. Similar processes are being established in many communities all over the world based on the principles of the healthy community movement (IOM, 1997b). Impressive improvements in health have been achieved when the entire community rallies behind a cause and focuses its collective efforts on making a positive change.

Each jurisdiction should also ensure that certain basic services are provided to all residents regardless of their ability to pay. The process used in each jurisdiction to determine these basic services should be coordinated by the jurisdictional health authority and should involve traditional leaders as well as residents of the broader community.

The committee does not recommend any particular health improvement benchmarking process. Each community and jurisdiction will need to determine what works best, given its unique circumstances and culture. U.S. federal agencies, particularly the Department of Health and Human Services, are encouraged to provide technical assistance to all communities that undertake such a process. Once Pacific communities have gained experience (good or bad) with health improvement benchmarking, they should be supported to act as sources of technical assistance to other communities.

Naturally, this process could prove to be politically problematic, especially when making decisions about appropriate referrals for off-island tertiary care or the continuation or expansion of a jurisdiction's hemodialysis program, for example. The committee encourages the jurisdictions to be clear about the process and the criteria to be used in making decisions. For example, Kosrae, in its *Healthy Kosrae 2000* report (Kosrae Department of Health Services, 1996), recognizes that its health care budget will decrease in the coming years and has begun to develop a list of what it considers to be essential services. In Kosrae, the following criteria are to be used to determine whether a service is essential: the effect of a service in decreasing morbidity and mortality and producing improved functional health, the bona fide needs of the Kosraen people and health providers (rather than the dictates of foreign donors), the availability of personnel resources and medical equipment, and the cost-effectiveness of the service.

Beyond its desire to have the communities themselves decide how well their health care systems are doing and what they will do to make improvements, the committee chose not to re-create or modify the benchmarks used in the University of Hawaii studies for a number of other practical and pragmatic reasons. The earlier studies involved months of on-site field work. The Institute of Medicine (IOM) contract with HRSA only called for very limited site visits lasting—at most—several days in each jurisdiction. This would mean that the jurisdictions would have to go to considerable effort to supply most of the needed information—a costly and time-consuming effort with limited direct utility.

Once the committee decided not to make use of the benchmarks, it did not evaluate them in any greater detail. However, the committee raises the following points for the jurisdictions to consider if they decide to re-create a similar set of benchmarks.

- An individual's decision about what is "adequate" would not constitute a reliable benchmark. As much as possible, groups should define and agree on common definitions, including what is "adequate" and "acceptable."

- The University of Hawaii benchmarks were used only to conduct a structural assessment of the health care system. As such, they do not provide procedural or outcomes information, such as health status.
- The criteria were designed to discern whether each Pacific Basin jurisdiction met minimally acceptable levels of health services and systems that might be expected of any health care system in rural America. This comparison does not seem entirely appropriate given the differences in social history, culture, finances, and the nature of the geographical isolation between rural America and the Pacific Basin. The committee believes that the benchmarks should be used to gauge how effectively health care services in the individual jurisdictions meet the needs of residents and that it should not develop comparisons that are not apt.
- Some aspects of the health care system are not adequately captured by the benchmarks. For example, no criterion or indicator covers the use of telemedicine or presence of private-sector health care services.
- The benchmarks are not weighted or prioritized in any way. The result is a "laundry list" of concerns, with equal importance given to having in place, for example, a geriatric program and a childhood immunization program.

Use Nongovernmental Community Organizations to Provide Health Services

Nongovernmental community organizations are often better situated to perform many activities not easily done by the government. They are often the best and sometimes the only effective means of outreach to the rural communities. However, they remain a potent and much underutilized force, particularly in the freely associated states.

The committee believes that nongovernmental community organizations should be enlisted to provide a variety of health-related activities not currently being provided by the jurisdiction's government, including health education and peer counseling, whenever possible. In some cases the organizations will need to receive initial financial support from the government to anchor themselves in the community and to fund training and volunteer development, but for the most part these organizations should be voluntary and charitable.

The committee sees the creation and support of these nongovernmental organizations as a vehicle that can be used to harness the diverse energies of the community. Such organizations will strengthen and support the islands' civil societies on a scale and in a manner that is appropriate to the jurisdictions and their cultures. The sharing of information about successful community-based programs and the creation of policies for replicating such programs in other jurisdictions are strongly encouraged.

Increase Community Involvement with Primary Care Sites

For all jurisdictions, when U.S. funds are involved, the committee recommends requiring community commitment and involvement in the delivery of care and the maintenance of primary care sites (dispensaries). Minimal requirements would be (1) donation of land from the community or some of its members, with a clear deed attesting to the donation; (2) contribution to the construction of the facility either in the form of materials or labor; (3) commitment of the community to maintaining the facilities; and (4) contribution to the salary of the person(s) serving as (a) community health aide(s).

More responsibility for nontechnical maintenance should be shifted to local communities. Communities should be required to enter into contracts with the government. In these contracts the communities should agree to meet certain initial and ongoing conditions in order for the government to continue to provide health services in the community. In-kind donations should consist of time and labor as well as or in place of monetary contributions. For example, the CNMI health department recently opened a satellite outreach center in the community of San Antonio near a teen center. Efforts to involve the community with the operation of the center should be actively pursued. These efforts could range from recruiting local teens to help staff the center to starting an advisory council that includes representation from nearby residents who could perform outreach to the growing immigrant population that lives near the center.

PROMOTE THE EDUCATION AND TRAINING OF THE HEALTH CARE WORKFORCE

The committee is gravely concerned about maintaining the skills and knowledge of the current health care workforce and strengthening the region's local human capacity. Crises similar to those that necessitated the establishment of the PBMOTP are destined to repeat themselves over and over again unless strategies to address the need for an adequate and well-trained workforce are not proactively developed and implemented. The committee therefore recommends several educational activities to address the present lack of adequate training opportunities available to the health care workforce in the U.S.-Associated Pacific Basin.

These individual activities should be based on a comprehensive workforce development and training plan established by each jurisdiction.

The plan should consider not only how to enhance and improve the skills of current health care providers but also how to train new providers, particularly women, to address shortages and natural loss through retirement and attrition. In general, the committee believes that education and training are best accomplished in settings close to the local population. This allows for greater cultural competence and acceptance of the health care practitioner and means that vital health care services will be provided in the region.

Activities should include but not be limited to the following: (1) improve and support basic education; (2) use distance-based learning, telemedicine, and electronic data libraries; (3) provide postgraduate continuing medical education programs; (4) sponsor training for dentists; (5) sponsor training for nurses; and (6) provide health administration and systems management training to the chief health administrator in each jurisdiction.

Improve and Support Basic Education

Currently, the primary and secondary educational systems throughout the region do not adequately provide students with the skills that they need to participate in the health care workforce. Fundamental educational reform is vitally needed. In the short-term, existing successful educational programs that prepare students to deal with the demands of higher education and that increase the applicability of coursework in secondary schools with regional health care institutions of higher education need to be replicated and nurtured. Programs aimed at increasing the education of women are particularly encouraged because of the likely improvements in health status that will result (Marshall and Marshall, 1983).

Alternative ways of financing education also need to be developed and supported. The committee encourages private organizations to provide scholarships and other financial support to students in the region who are interested in pursuing careers in health care.

Use Distance-Based Learning, Telemedicine, and Electronic Data Libraries

The use of telecommunications for clinical consultation—whether through existing technologies such as shortwave radios, telephones, and the Internet or through more advanced technologies—is to be encouraged and supported. Ways to network or combine health telecommunications networks with other markets, such as education and private business, to decrease costs should be explored. Whenever possible common hardware and software standards should be used to avoid compatibility problems between and among sites.

The use of existing instructional materials in an interactive CD-ROM format and the adaptation of such materials to the unique aspects of the region should be explored. The development of such instructional materials by institutions of higher education in the region should be supported by the U.S. federal government.

In view of the serious practical limitations of the PEACESAT satellite system reported during the committee's site visits and workshop in Saipan, Pacific Basin jurisdictions and the institutions of higher education in the region should continue to explore more reliable means of communication. The establishment of educational links within the region and with other areas of the world should be seriously explored.

Provide Postgraduate and Continuing Medical Education Programs

Continuing medical education (CME) must be required and incorporated into the health care workforce training plan for the entire region. The committee is particularly concerned that the graduates of the Pacific Basin Medical Officers Training Program (PBMOTP) receive CME to improve and maintain their clinical skills and knowledge.

Several means of providing ongoing CME exist. Courses can be provided through the Pacific Basin Medical Association and, where available, through an individual jurisdiction's medical or professional associations. Whenever possible, training should be done within the jurisdiction, perhaps through periodic visits by other physicians and health care professionals who can provide personal support while offering consultative clinics and other training opportunities. The Archstone Foundation provides funding for the Medical Graduate Support Program, which allows a former PBMOTP faculty member to travel throughout the region to provide in-service training. Some additional in-service training has been provided though the Regional Training and Research Centre in Reproductive Health based at the Fiji School of Medicine, the John A. Burns School of Medicine (JABSOM) at the University of Hawaii, and teams of practitioners from the Tripler Army Medical Center (the missions of both JABSOM and Tripler, in fact, call for sending health care providers to the region). The departments of health of the various jurisdictions should work together to better plan and coordinate visits from such consultants and voluntary medical teams.

The committee foresees the continued need for U.S. Public Health Service and National Health Service Corps personnel in the region, particularly in providing training to the existing workforce. Their assignment should reflect the needs identified by each jurisdiction in its health care workforce development plan. Each jurisdiction should also develop a description of the assignment that accurately describes the context in which the volunteer will be working and exactly what the volunteer's role will be. Volunteers should also be allowed and encouraged to provide on-site training and quality assurance evaluations in lieu of clinical care. The committee draws particular attention to the opportunity of recruiting and tailoring the education of National Health Service Corp-sponsored medical students, nurse practitioners, and nurse anesthetists at the Uniformed Services University of the Health Sciences. These students have already made a commitment of service to the country and several have completed training programs in the region.

Pacific Basin educational institutions and other organizations serving the region, such as WHO and the SPC, should collaboratively develop additional training programs and CME courses for all types of health care providers. The training should be appropriate for the region but should emphasize primary care and community health.

Formal postgraduate education should be conducted at a regional training center, preferably an existing one. The availability of such programs should be open to all practitioners in the region. The role of the U.S. federal government should be directed toward capacity building and financial assistance.

Sponsor Training for Dentists

The committee is greatly concerned about the dearth of dentists currently practicing in the region. Many of the current dental practitioners are expatriates or are nearing retirement. The committee therefore strongly recommends that the U.S. federal government and local jurisdictions sponsor dental training immediately. A dental officer program based on the PBMOTP model was developed by PIHOA in consultation with the University of Kentucky School of Dentistry in the early 1990s (University of Kentucky, 1993). Although it was never funded, the committee believes that the basic model, which provides a graduated career ladder approach to training with substantial clinical work, is a sound one. Fortunately, the dental program at the Fiji School of Medicine has recently been reorganized using a similar model and would provide a good base for some of this much needed dental training to occur. In recent years improvements in distance-based education have also made it possible for the current and potential dental workforce to be trained in their own jurisdictions. This has the benefit of allowing dental practitioners to continue to provide their much needed services on-island and should be seriously explored and supported.

Sponsor Training for Nurses

Nurses play an important role in the region's health care delivery system. Yet in nearly every jurisdiction, officials reported having a nursing shortage. To address this situation, the committee recommends the following steps be taken. Training for nurses and for individuals in various allied health fields should continue to take place, as it does now, in several institutions of higher education located throughout the region. The committee is particularly concerned, however, about the high dropout and low graduation rates among individuals in every nursing program in the region. Regional nursing programs are encouraged to work together to evaluate the reasons for these problems and design new approaches to address them.

One approach that has been successfully implemented in some jurisdictions and that could be pursued further is the inclusion of clinical experience in community settings and the use of curricula that emphasize culturally appropriate primary care. Other approaches include the sharing of faculty members, the use of cooperative efforts to provide distance-based education, upgrading curriculum to the bachelor's level, and development of continuing nursing education programs for existing nurse personnel at all levels, which is viewed as essential to strengthen inpatient, outpatient, and community health care services. Some of these efforts have already begun, but they need to be sustained. HRSA's Bureau of Health Professions and its Division of Nursing should provide funds for nurse traineeships and other special training programs identified by the regional nursing programs.

Provide Health Administration and Systems Management Training to the Chief Health Administrator

(See earlier recommendation on reforming health care facility management.)

Bibliography

Abraham, I. CNMI's accomplishments—past, present, and future. Remarks at the Institute of Medicine Workshop on Health Care Services in the U.S.-Associated Pacific Basin held on Saipan, Commonwealth of the Northern Mariana Islands, April 18, 1997.

Ahlburg, D.A. Demographic and social change in the island nations of the Pacific. *In Asia-Pacific Population Research Reports*, Number 7. Honolulu: East-West Center, Program on Population, February, 1996.

Arvis, C. Presentation at the meeting of the Institute of Medicine Committee on Health Care Services in the U.S.-Associated Pacific Basin, Washington, D.C., January 10, 1997.

Bank of Hawaii. *Palau Economic Report. An Economic Assessment of the Republic of Palau, September 1994.* [WWW document]. URL http://www.boh.com/econ/411palaer.html (accessed December 13, 1996), 1994.

Bank of Hawaii. *Federated States of Micronesia Economic Report.* [WWW document]. URL http://www.boh.com/econ/pacific/fsmaer.html (accessed November 10, 1997), Autumn, 1995a.

Bank of Hawaii. *Guam Economic Report.* [WWW document]. URL http://www.boh.com/econ/pacific/gmaer.html (accessed November 10, 1997), Summer, 1995b.

Bank of Hawaii. *Republic of Marshall Islands Economic Report, Winter 1995–1996.* [WWW document]. URL http://www.boh.com/econ/411rmiaer.html (accessed December 30, 1997), 1996.

Bank of Hawaii. *American Samoa Economic Report.* [WWW document]. URL http://www.boh.com/econ/pacific/as/index.html (accessed August 15, 1997), April, 1997.

Bell, T. Presentation given at the meeting of the Institute of Medicine Committee on Health Care Services in the U.S.-Associated Pacific Basin, Washington, D.C., January 10, 1997.

Bice, S., G. Dever, L. Mukaida, S. Norton, and J. Samisoni. Telemedicine and telehealth in the Pacific Islands region: A survey of applications, experiments, and issues, *Proceedings of the Pacific Telecommunications Conference* pp. 574–581. [WWW document]. URL http://obake.peacesat. Hawaii.edu/info/papers/telemed4.htm (accessed February 19, 1997), 1996.

Brazeal, Aurelia E. (U.S. Department of State, Bureau of East Asian and Pacific Affairs). U.S. relations with the freely associated states. Testimony presented before the Subcommittee for Asia and the Pacific, House Committee on International Relations, Washington, D.C., September 25, 1996 [WWW document]. URL http://www.state.gov/www/regions/eap/brazeal.html (accessed December 30, 1996), 1996.

Brewis, A., P. Schoeffel-Meleisea, H. Mavoa, and K. Maconaghie. Gender and non-communicable diseases in the Pacific. *Pacific Health Dialog* 3(1):107–112, 1996.

Bruss, J. *Annual Public Health Progress Report.* Commonwealth of the Northern Mariana Islands, Department of Public Health, Division of Public Health, Saipan, CNMI: CNMI Dept. of Public Health, 1996.

Centers for Disease Control and Prevention. *Australia and the South Pacific* [WWW document]. URL http://www.cdc.gov/travel/austspac.htm (accessed March 27, 1997), 1997.

Commonwealth of the Northern Mariana Islands, Department of Public Health and Environmental Services. Costs of Health/Medical Services Provided to Micronesians. Saipan: Commonwealth of Northern Mariana Islands Department of Public Health, January, 1997.

Dever, G. Information and Current Mailing Addresses for Pacific Basin Medical Officer Training Program Graduates. Kolonia: Federated States of Micronesia Human Resource Development Center, 1997a.

Dever, G. Remarks at the Institute of Medicine Committee on Health Care Services in the U.S.-Associated Pacific Basin, workshop in Saipan, Commonwealth of Northern Mariana Islands, April 18, 1997.

Diaz, A. The health crisis in the U.S.-Associated Pacific Islands: Moving forward. *Pacific Health Dialog* 4(1):116–129, 1997.

Epstein, L. Linguistically and Culturally Appropriate Services for Asian Americans and Pacific Islanders: Implications for HP 2000/2010. Paper presented at the Asian and Pacific Islander American Health Forum, San Francisco, September 12–14, 1997.

Federated States of Micronesia. *Healthy Nation 2000.* Palikir: Government of the Federated States of Micronesia, 1996.

Federated States of Micronesia. *Second National Development Plan, 1992–1996.* Palikir: Government of the Federated States of Micronesia, November 13, 1991.

Flear, J. The evolution of community health training at the PBMOTP. *Pacific Health Dialog* 4(1):198–202, 1997a.
Flear, J. Good Intentions: Good Enough? Paper commissioned by the Institute of Medicine Committee on Health Care Services in the U.S.-Associated Pacific Basin, 1997b.
Fochtman, M. , C. Allen, and R. Gurusamy. Distance education for health workers in Micronesia. *Pacific Health Dialog* 4(1):203–206, 1997.
GAO (General Accounting Office). *American Samoa: Inadequate Management and Oversight Contribute to Financial Problems.* Report GAO/NSIAD-92-64. Washington, D.C.: General Accounting Office, 1992.
GAO. *Veterans' Benefits: Availability of Benefits in American Samoa.* Report GAO/HRD-93-16. Washington, D.C.: General Accounting Office, 1993.
Guam Health Planning and Development Agency. *Guam Health Plan: 2001 and Beyond.* Agana, Guam: Guam Planning and Development Agency, June, 1996.
Haddock, R.L. Cancer in Guam: A review of death certificates from 1971–1995. *Pacific Health Dialog* 4(1):66–75, 1997.
Haddock, R.L., J.H. Hoffman, and W.R. Williams. Betel nut chewing on Guam. *Fiji Medical Journal* 9:139–145, 1981.
Head, P. Is it really a man's world? *Pacific Health Dialog* 4(1):5–7, 1997.
Hezel, F.X. *Strangers in Their Own Land: A Century of Colonial Rule in the Caroline and Marshall Islands.* Honolulu: University of Hawaii Press, 1995.
Hezel, F.X., E.Q. Petteys, and D. Chang. Sustainable Human Development in the Federated States of Micronesia. Paper prepared for the United Nations Development Programme, March 19, 1997.
Howard, J., P. Heotis, W. Scott, and W. Adams. *Medical Status of Marshallese Accidentally Exposed to 1954 Bravo Fallout Radiation; January 1988 Through December 1991.* Washington, D.C.: U.S. Department of Energy, July 1995.
HRSA (Health Resources and Services Agency). Draft Report of the HRSA Pacific Basin Intra-Agency Workgroup. U.S. Department of Health and Human Services, July 25, 1996.
IOM (Institute of Medicine). *Dental Education at the Crossroads: Challenges and Change.* M. Field, ed. Washington, D.C.: National Academy Press, 1995.
IOM. *Primary Care: America's Health in a New Era.* M. Donaldson, K. Yordy, K. Lohr, and N. Vanselow, eds. Washington, D.C.: National Academy Press, 1996.
IOM. *America's Vital Interest in Global Health.* C.P. Howson, ed. Washington, D.C.: National Academy Press, 1997a.
IOM. *Improving Health in the Community: A Role for Performance Monitoring.* J. Durch, L. Bailey, and M. Stoto, eds. Washington, D.C.: National Academy Press, 1997b.

Kosrae Department of Health Services. *Healthy Kosrae 2000: A Health Services Plan for the State of Kosrae.* Kosrae Department of Health Services; September, 1996.

Larin, J., T. Gulick, and L. Pederson. *Medical Referral Improvement Project for American Samoa.* Honolulu: Asian/Pacific Research Foundation, 1994.

Levin, M.J. *Micronesian Migrants to Guam and the Commonwealth of the Northern Mariana Islands: A Study of the Impact of the Compact of Free Association.* Washington, D.C.: U.S. Bureau of the Census, April 20, 1996.

Lin-Fu, J. Ethnocultural Barriers to Health Care: A Major Problem for Asian and Pacific Islander Americans. *Asian American and Pacific Islander Journal of Health* 4(2):290–298, 1994.

Marshall, M. *Weekend Warriors: Alcohol in Micronesian Culture.* World Ethnology Series. Palo Alto: Mayfield Publishing Co., 1979.

Marshall, L. and Marshall, M. Education of Women and Family Size in Two Micronesian Communities. *Micronesia* 18(1):1–21, 1983.

Marshall, M. Young men's work: alcohol use in the contemporary Pacific. In A.B. Robillard and A.J. Marsella, eds. *Contemporary Issues in Mental Health Research in the Pacific Islands.* Honolulu: University of Hawaii, 1987.

Marshall, M. The second fatal impact: Cigarette smoking, chronic disease, and the epidemiological transition in Oceania. *Social Science Medicine* 33(12):1327–1342, 1991.

Medical Graduate Support Program. *Dispensaries in the Federated States of Micronesia (Chuuk, Kosrae, Pohnpei, and Yap).* Kolonia, Federated States of Micronesia: Medical Graduate Support Program, 1997.

Micronesian Seminar. *Alcohol and Drug Use in the Federated States of Micronesia: An Assessment of the Problem with Implications for Prevention and Treatment.* Report prepared for the Center for Substance Abuse Treatment, Substance Abuse and Mental Health Services Administration, U.S. Department of Health and Human Services. Pohnpei, Federated States of Micronesia: Micronesian Seminar, March 1997.

National Cancer Institute. Estimated Exposures and Thyroid Doses Received by the American People from Iodine-131 in Fallout Following Nevada Atmospheric Nuclear Bomb Tests: A Report from the National Cancer Institute. NIH Publication No. 97-4264. Washington, D.C.: U.S. Department of Health and Human Services, 1997.

O'Leary, M. Health data systems in Micronesia: Past and future. *Pacific Health Dialog* 2(1):126–132, 1995.

Pacific Island Health Officers Association (PIHOA). The Strategic Five Year Plan of the Pacific Island Health Officers Association, 1996–2000. Honolulu: Pacific Island Health Officers Association Executive Office, January 1996.

PIHOA. Health Data Matrix. Update of PIHOA, 1994 Data Matrix prepared for the Institute of Medicine Committee on Health Care Services in the U.S.-

Associated Pacific Basin. Honolulu: Pacific Island Health Officers Association, 1997.

Palafox, N.A., D.B. Johnson, A.R. Katz, J.S. Minami, and K. Brian. Site-specific cancer incidence in the Republic of the Marshall Islands. *Cancer*, in press.

Pretrick, E. The PBMOTP—Towards an inspirational accomplishment. *Pacific Health Dialog* 4(1):94–98, 1997.

Pryor, M., J. Pryor, J. Manning, S. Manning, R. Rudoy, J. Stewart, G. Dever, and S. Stool. Vitamin A deficiency and otitis media in Chuuk State, Micronesia. *Pacific Health Dialog* 1(1):6–12, 1994.

Republic of the Marshall Islands (RMI). *Marshall Islands Statistical Abstract: 1988.* Majuro, Republic of the Marshall Islands: Office of Planning and Statistics, 1988.

Republic of the Marshall Islands. *National Population Policy.* Majuro, Republic of the Marshall Islands: Office of Planning and Statistics, 1990.

Ruze, P. Presentation at the Institute of Medicine's Committee on Health Services in the Pacific Basin meeting, Washington, D.C., August 22, 1997.

Sacks, O. *The Island of the Colorblind.* New York: Alfred A. Knopf, 1996.

Samo, M. *External Funding Contribution to Health Care Development in the Federated States of Micronesia.* Kolonia, Federated States of Micronesia: Micronesian Seminar, 1997.

Sarda, R. and G. Harrison. Epidemiology in the Pacific. *Pacific Health Dialog* 2(2):6–14, 1995.

Simon, S., ed. The consequences of nuclear testing in the Marshall Islands. *Health Physics* 73(1; Special Ed.), July, 1997.

Takahashi, T., K.R. Trott, K. Fujimori, S.L. Simon, H. Ohtomo, N. Nakashima, K. Takaya, N. Kimura, S. Satomi, and M.J. Schoemaker. An investigation into the prevalence of thyroid disease on Kwajalein Atoll, Marshall Islands. *Health Physics* 73(1):199–213, 1997.

U.S. Bureau of the Census. *Population Estimates Program* [WWW document]. URL http://www.census.gov/population/estimates (accessed November 17, 1997), 1997a.

U.S. Bureau of the Census. International Programs Center. International Database [WWW document]. URL http://www.census.gov/cgi-bin/ipc/idbsum (accessed July 8, 1997), 1997b.

U.S. Bureau of the Census. *Federal and State Expenditures by State for Fiscal Year 1996.* Washington, D.C.: U.S. Government Printing Office, 1997.

U.S. Bureau of the Census. 1990 U.S. Census Data Summary Tape File 1C. [WWW document] URL http://venus.census.gov/cdrom/lookup (accessed 12/21/97), 1990.

U.S. Department of Health and Human Services. *Healthy People 2000 Review 1995–96.* Publication No. (PHS) 96-1256. Hyattsville, Md.: U.S. Department of Health and Human Services, 1996.

U.S. Department of Health and Human Services (USDHHS), Centers for Disease Control and Prevention, National Center for Health Statistics.

Health, United States 1996–97 and Injury Chartbook. Hyattsville, Md.: U.S. Department of Health and Human Services, 1997.
U.S. Department of the Interior, Office of Insular Affairs. *The Impact of the Compacts of Free Association on the United States Territories and Commonwealths and on the State of Hawaii.* A report to the U.S. Congress, Washington, D.C.: U.S. Department of the Interior, September 1996a.
U.S. Department of the Interior, Office of Insular Affairs. *Report on the State of the Islands.* Washington, D.C.: U.S. Department of the Interior, August 1996b.
U.S. Department of the Interior, Office of Insular Affairs. Fact Sheets, United States Insular Areas and Freely-Associated States. [WWW document.] URL http://www.doi.gov/oia/oiafacts.html (accessed December 16, 1997), 1997.
U.S. Department of State, Office of Public Communication. *Background Notes: Marshall Islands*, Vol. 4, No. 3. Washington, D.C.: U.S. Department of State Bureau of Public Affairs, February 1994.
University of Hawaii. *Health Services in U.S. Pacific Island Jurisdictions.* Honolulu: University of Hawaii at Manoa, School of Public Health, 1984.
University of Hawaii. *A Reevaluation of Health Services in U.S.-Associated Pacific Island Jurisdictions.* Honolulu: University of Hawaii at Manoa, School of Public Health, 1989.
University of Kentucky College of Dentistry. *Technical and Financial Proposals for the Development of a Pacific Island Dental Officer Program.* Lexington, Kentucky: University of Kentucky College of Dentistry, 1993.
University of the South Pacific. *South Pacific Web Atlas* [WWW Document]. URL http://www.usp.ac.fj/~gisunit/pacatlas/ATLAS.htm (accessed August 15, 1997), 1997.
Ventura, S.J., K.D. Peters, J.A. Martin, and J.D. Maurer. Births and deaths: United States, 1996. *Monthly Vital Statistics Report,* Vol. 46, Issue 1, Supplement 2. Hyattsville, Md.: National Center for Health Statistics, September 11, 1997.
Workman, A. and D. Quintana. Epidemiology as labeling: Neurological diseases and stigma on Guam. *Journal of Micronesian Studies* 4(1):47–69.
World Bank. *World Development Report 1993: Investing in Health.* Oxford, England: Oxford University Press, 1993.
World Bank. *Health Priorities and Options in the World Bank's Pacific Member Countries.* Report 11620-EAP. Washington, D.C.: World Bank, 1994.
WHO (World Health Organization). Obesity Epidemic Puts Millions at Risk from Related Diseases. WHO Press Release. Geneva: World Health Organization, June 12, 1997.
WHO. Most Recent Values for WHO Global Health for All Indicators, Marshall Islands. [WWW document.] URL http://www.who.ch/programmes/hst/hsp/a/countrys/mar2.htm (accessed 12/16/97), August 1996.
WHO. The Yanuca Island Declaration on Health in the Pacific in the 21st Century. Agreement adopted at the Conference of Ministers of Health of the Pacific Islands, Yanuca Island, Fiji, March 10, 1995.

WPHNet. *Infoline*. Kolonia, Federated States of Micronesia: Pacific Basin Medical Association, (1)2:4, September 6, 1997.

Yaingeluo, J. Environmental risks for respiratory, diarrhoeal and skin diseases in six Pohnpeian villages. *Pacific Health Dialog* 3(2):187–193, 1997.

Youth to Youth in Health. *Program Report 1995–96*. Majuro, Republic of the Marshall Islands: Youth to Youth in Health, 1997.

Appendix A

Committee Biographies

ROBERT S. LAWRENCE, M.D., Committee Chair, is Associate Dean for Professional Education and Programs and Professor of Health Policy and Management at the Johns Hopkins School of Hygiene and Public Health and Professor of Medicine at the Johns Hopkins School of Medicine. He is a graduate of Harvard College and Harvard Medical School and trained in internal medicine at the Massachusetts General Hospital in Boston, Massachusetts. He served for three years as an Epidemic Intelligence Service Officer at the Centers for Disease Control, U.S. Public Health Service. He served as Chief of Medicine at the Cambridge Hospital and the Director of the Division of Primary Care and the Charles S. Davidson Associate Professor of Medicine at the Harvard Medical School until 1991. From 1991 to 1995 he was Director of Health Sciences at the Rockefeller Foundation. From 1984 to 1989 Dr. Lawrence chaired the U.S. Preventive Services Task Force of the Department of Health and Human Services and served on the successor Preventive Services Task Force from 1990 to 1995. Dr. Lawrence is a fellow of the American College of Physicians and the American College of Preventive Medicine and is a member of the Institute of Medicine, the Association of Teachers of Preventive Medicine, the Society of General Internal Medicine, and the American Public Health Association.

DYANNE D. AFFONSO, R.N., Ph.D., is Dean and Professor at the Nell Hodgson Woodruff School of Nursing, Emory University, and Associate Professor in the Women's and Children's Division of the School of Public Health. Previously, she was a faculty member at the School of Nursing of the University of California at San Francisco and the College of Nursing, University of Arizona. She is a leading authority in a variety of maternal-child health topics and plays an active role in recruitment and mentoring of minority women in biomedical careers. Her recent publications focus on health issues for women from ethnically diverse and rural backgrounds, with an emphasis on Asian-Pacific Islanders. She is a member of the Institute of Medicine.

CAROLYNE K. DAVIS, R.N., Ph.D., is a national and international health care adviser to Ernst & Young. She received a B.S. in nursing from the Johns Hopkins University, and an M.S. in nursing and a Ph.D. in higher education administration from Syracuse University. She has been Chair of the baccalaureate nursing program at Syracuse University and held many positions at the University of Michigan, Ann Arbor, including Dean of the School of Nursing, Professor of both nursing and education, and Associate Vice President for Academic Affairs. Following this, she became the fourth Administrator of the Health Care Financing Administration and held that position from 1981 to 1985. As Administrator, Dr. Davis oversaw the functions of the Medicare and Medicaid programs, which finance health care services for 54 million poor, elderly, and disabled Americans. She is a member of the editorial board of the journal *Nursing Economics* and has more than 100 publications on a wide variety of issues concerning the health care system. Dr. Davis has received many honorary degrees and alumna awards, is on the board of directors or is a member of the board of several corporations, and is a member of the Institute of Medicine.

WILLIAM H.J. HAFFNER, M.D., Captain, U.S. Public Health Service (PHS), has been Chair of the Department of Obstetrics and Gynecology at the Uniformed Services University of the Health Sciences since September 1992. He began his PHS career with the Navajo Area Indian Health Service in 1971, where he served as Head of the Department of Obstetrics and Gynecology until 1980 and then as the Obstetrics and Gynecology Consultant for the entire Indian Health Service from 1980 through 1994. Upon his transfer to the National Capital Area in 1981, he has served the Indian Health Service in a variety of consultative roles, and he was appointed the Chief Medical Officer of the PHS for two terms from 1990 through 1994. Dr. Haffner maintains an active clinical practice in obstetrics and gynecology at the National Naval Medical Center and the Walter Reed Army Medical Center. Dr. Haffner is a graduate of the George Washington University School of Medicine, and completed his residency training in Obstetrics and Gynecology at the Columbia Presbyterian Medical Center, including 2 years as Chief of Residents. He is certified by the American Board of Obstetrics and Gynecology.

GLEN E. HAYDON, M.S., F.A.C.H.E., recently retired from the position of President of Mercy International Health Services. Mercy International is sponsored by the Religious Sisters of Mercy and is organized to provide leadership and other training for hospital and clinic personnel and counsel to individual hospital, ministry of health, and other governmental leaders. Prior to serving in this position Mr. Haydon had served in various senior hospital administrative positions for the Sisters spanning 26 years. At one point he was responsible for the administration of seven rural hospitals in northern Iowa ranging in size from 35 to 90 beds. Mr. Haydon has served as Chairman of Iowa's Certificate of Need Committee, a member of the Health Systems Agency Board of Directors, and Chairman of the Iowa Health Plan Committee. He also held various positions with the Iowa Hospital Association and was Chairman of the Iowa Commission on Aging for 6 years. During his tenure he served as an Iowa Delegate to the White House Conference on Aging. He has directed medical and relief operations in a number of civil wars and natural disasters in Africa, Bangladesh, and Peru. To date, Mr. Haydon has provided counsel to health care delivery workers in 29 countries. Mr. Haydon has received awards from the International Committee of the Red Cross the White House, and the American Red Cross, for his work in developing countries. Mercy International Health Services also honored him with its highest award for service, recognizing particularly his zeal to assist health care workers in the less developed world.

FRANCIS X. HEZEL, S.J., is a Jesuit priest who has worked in Micronesia since 1963, when he first came to Chuuk to teach at Xavier High School. He is Director of the Micronesian Seminar, a pastoral-research institute that has been heavily involved in public education for 25 years. In that capacity he has written articles on many phases of development in the region, sponsored numerous conferences, and authored several books and numerous articles on local history. Since 1992 he has been residing on Pohnpei and is now the regional superior of the Jesuits of Micronesia.

AGNES MANGLONA McPHETRES, M.A. Ed., has been President of the Northern Marianas College in Saipan since 1983. From 1978 to 1983 she was the Superintendent of Education for the Commonwealth of Northern Mariana Islands (CNMI). Ms. McPhetres was actively involved in the educational system of the Trust Territory of the Pacific Islands and served in the Office of Transition when the Northern Marianas Islands achieved commonwealth status. She is a charter member of the Pacific Postsecondary Education Council, on which she served as chair from 1983 to 1985, and from 1993 to the present. She was also chair of the CNMI Humanities Council from 1990 to 1992. She is a member of the Accrediting Commission for Community and Junior Colleges, Western Association of Schools and Colleges, and from 1980 to 1984 served on the Commission of Federal Laws Applicable to CNMI.

HON. TOSIWO NAKAYAMA is currently advisor to the Pacific Islands Development Bank, as well as Vice President of Governmental Affairs of the Bank of Guam. He comes to that position having served as the Federated States of Micronesia's first President, from 1979 to 1987. Former President Nakayama was the President of the Micronesia Constitutional Convention in 1975 and was a member of the Congress of Micronesia from 1965 to 1979, capping a distinguished political career that began in the Truk District Congress in 1959.

He served as advisor to the U.S. Mission to the United Nations for Trusteeship Council in 1962, in 1972, and from 1979 to 1986. He attended the University of Hawaii and in 1982 received its highest alumnus award, the Rainbow Award, and the East-West Center's Distinguished Alumnus award for 1984. The University of Guam awarded him an honorary Doctor of Laws degree in 1987.

PAUL W. NANNIS, M.S.W., is Commissioner of Health, City of Milwaukee, Wisconsin, a position he has held since 1988, with the exception of working for a half year for the Robert Wood Johnson Foundation as a Senior Program Officer in 1995 and 1996. He was Executive Director of the 16th Street Community Health Center from 1979 to 1988. The 16th Street Community Health Center is a federally funded, multiservice health center caring for an ethnically mixed population in Milwaukee. Paul attended the Health Executives Development Program at Cornell University in 1985 and the Public Health Leadership Institute sponsored by the Centers for Disease Control and the University of California-Santa Cruz in 1992 and 1993. He was President of the Board of Directors of the U.S. Conference of Local Health Officers for two terms between 1991 and 1994, and he is currently a member of the Executive Committee of the National Association of County and City Health Officials. Paul also served on the Institute of Medicine Committee on the Future of Primary Care from 1994 to 1996.

TERENCE A. ROGERS, Ph.D., served as Dean of the John A. Burns School of Medicine at the University of Hawaii from 1971 to 1988. He was the Director of the Hawaii State Hospital from 1991 to 1993 and Counselor for Congressional Relations at the East-West Center in Honolulu from 1989 to 1991. He was Professor of Physiology at the John A. Burns School of Medicine from 1963 to 1989. Dr. Rogers took a sabbatical at Kyoto Prefectural University of Medicine in Japan from 1969 to 1970 and served on President Carter's Commission on World Hunger in Washington, D.C. from 1978 to 1980. Dr. Rogers has also served on the faculties of Stanford University School of Medicine and the University of Rochester School of Medicine.

DAVID N. SUNDWALL, M.D., is President of the American Clinical Laboratory Association (ACLA), which represents the leading national, regional, and local independent clinical laboratories. Prior to joining ACLA, Dr. Sundwall was Vice President and Medical Director of American Healthcare Systems, the largest coalition of not-for-profit hospital systems in the United States. Previous federal policy positions include Administrator of the Health Resources and Service Administration, U.S. Public Health Service, and Health Staff Director of the Labor and Human Resources Committee in the U.S. Senate. Dr. Sundwall is a graduate of the University of Utah (B.A. and M.D.) and completed his internship and residency in the Harvard University Family Practice Medicine Program. He is board-certified in internal medicine and family practice.

Appendix B

Saipan Workshop Agenda

INSTITUTE OF MEDICINE
Committee on Health Care Services in the U.S.-Associated Pacific Basin

SAIPAN MEETING AGENDA
Hyatt Regency, Gilligan's
Friday, April 18, 1997

(The meeting began 2 hours later than originally planned to allow travelers coming from Guam who had been delayed by Typhoon Isa to arrive.)

11:00 a.m. Prayer: **Father Roger Tenorio**

11:05 Introductory Remarks:
Robert Lawrence, IOM Committee Chair
Eliuel Pretrick, PIHOA Vice President

11:15 Opening Remarks: **Acting Governor Jesse Borja,** *CNMI*

11:25 Introduction of the Institute of Medicine mission and of individual committee members

11:30 **Comments from individual health ministers:** The ministers or representatives have been asked to list the three accomplishments that they have been most proud of in the past 10 years and three of the most important issues that they plan to address over the next 10 years.

1:00 p.m. Lunch—*Sponsored by the Governor's Office and the Department of Public Health and Environmental Services*

2:00	Social and Cultural Service Delivery Issues in Health Care: **Joe Flear**, *Medical Graduate Support Program*
	(Responses from individual health ministers)
2:30	Health Professions Training and Workforce Issues: Medical Officers/Physicians—Postgraduate Opportunities **Greg Dever**, *Director, Micronesian Human Resources Development Council,* and **Vita Skilling** *and* **Anamarie Akapito**, *PBMOTP graduates,* **Jimione Samisoni** *and* **Wame Baravilala**, *Fiji School of Medicine*
4:00	Basic Education Issues in the Pacific Primary Education—**Fran Hezel**, *IOM Committee Member, Director, Micronesian Seminar* Postsecondary Education—**Agnes McPhetres**, *IOM Committee Member, President, Northern Marianas College*
4:50	Recap of Key Points from the Day's Session: **Eliuel Pretrick** and **Robert Lawrence**
5:00	Adjourn
6:00	**"Hafa Adai" Hospitality**—Luau at Susupe Beach Park: *Sponsored by Marianas Visitors Bureau and Northern Marianas College*

Saturday, April 19, 1997

8:00 a.m.	Recap of Yesterday's Themes: **Robert Lawrence** and **Eliuel Pretrick**
8:15	Community Health Improvement: Block Grants and Community Empowerment Transforming Indian Health: **William Haffner**, *IOM Committee Member* Hawaii's Experience with Improving Prenatal Care: **June Shibuya**
9:00	Small Group Discussions
10:30	Promoting Privatization: **Isamu Abraham**, *Secretary, CNMI Department of Public Health and Environmental Services,* and **Susan Schwartz**, *President of CNMI-CHC Volunteers*

APPENDIX B 91

11:00	Sources of International Health Aid: **Marcus Samo**, *FSM Department of Health*
12:15	Lunch—*Sponsored by the Bank of Guam*
1:30 p.m.	Telemedicine and Telehealth in the Pacific Basin: **Greg Dever**, *Director, Micronesian Human Resources Development Council* **Maureen Fochtman**, *Professor, University of Guam, School of Nursing*
2:00	Mental Health and Substance Abuse Issues: **Elena Scragg**, *Guam Department of Mental Health* **Joe Villagomez**, *CNMI Division of Mental Health*
3:00	Increasing the IOM Report's Impact at the Local Level
4:00	Summarizing Themes and Recommendations: **Eliuel Pretrick** and **Robert Lawrence**
4:30	Adjourn
4:45	Closed Session: IOM Committee Meeting
7:00	Reception—Pacific Island Club: *Sponsored by Saipan Health Clinic and FHP-Health Care*

Support for this meeting came from the U.S. Department of Health and Human Services, Health Resources and Services Administration, the U.S. Department of Interior, and the Archstone Foundation.

SAIPAN WORKSHOP ATTENDEES

IOM Committee Members
Carolyne Davis
William Haffner
Glen Haydon
Fran Hezel
Robert Lawrence
Agnes McPhetres
Tosiwo Nakayama
Paul Nannis
Terence Rogers
David Sundwall

IOM Staff Members
Jill Feasley
Annice Hirt

PIHOA Members
Isamu Abraham, *CNMI*
Asher Asher, *Kosrae*
Aminis David, *Pohnpei*
Caleb Otto, *Palau*
Eliuel Pretrick, *FSM*
Elena Scragg, *Guam*

Joseph Tufa, *American Samoa*
Sanphy William, *Chuuk*
Andrew Yatilman, *Yap*

PIHOA Staff
Jean Paul Chaine
Roylinne Wada

Regional Participants
Anamarie Akapito, *Chuuk*
Wame Barivilala, *Fiji*
Jefferson Benjamin, *FSM*
Gregorio Calvo, *CNMI*
Karen Cruz, *Guam*
Greg Dever, *Pohnpei*
Joseph Flear, *Pohnpei*
Maureen Fochtman, *Guam*
Steve Kuartei, *Palau*
Jan Pryor, *Pohnpei*
Helen Ripple, *Guam*
Josephine Sablan-Hall, *CNMI*
Jimi Samisoni, *Fiji*
Marcus Samo, *FSM*

Susan Schwartz, *CNMI*
June Shibuya, *Hawaii*
Vita Skilling, *Kosrae*
Joe Villagomez, Jr., *CNMI*

Other U.S. Participants
Anne Chang, U.S. Health Resources and Services Administration
Mary Ellen Courtright, Archstone Foundation
Bruce Grant, Substance Abuse and Mental Health Services Administration
Joe Iser, U.S. Department of Health and Human Services
Darla Knoblock, Department of Interior
Gordon Soares, U.S. Health Resources and Services Administration
Joe Webb, Centers for Disease Control and Prevention

Appendix C

Organization of Insitute of Medicine Site Visits to the Pacific Basin, 1997

Island	Site Visitor	Date of Site Visit
American Samoa	Jill Feasley, Terence Rogers	April 8–11
Chuuk	Glen Haydon, Robert Lawrence	April 15–17
Commonwealth of the Northern Mariana Islands	Jill Feasley, Francis Hezel, Agnes McPhetres	April 16 and 17
Fiji	Jill Feasley	April 5–8
Guam	(a) Jill Feasley (b) Carolyne Davis, Tosiwo Nakayama, and Paul Nannis (c) Annice Hirt	(a) April 14 and 15 (b) April 17 (c) April 17, April 21–25
Hawaii	Glen Haydon, Annice Hirt, David Sundwall	April 10
Kosrae	David Sundwall	April 12–15
Republic of the Marshall Islands	Dyanne Affonso	July 8–11
Republic of Palau	Carolyne Davis	April 13–16
Pohnpei	(a) Glen Haydon (b) William Haffner, David Sundwall	April 12–15 April 15–17
Yap	Annice Hirt, Paul Nannis	April 13–16

D

Assessments of Individual Jurisdictions' Health Care Services

Although the six jurisdictions of the U.S.-Associated Pacific Basin are often grouped together into one geographic category, as demonstrated in Chapter 2, such groupings tend to downplay the very different situations that exist in each unique jurisdiction. This appendix assesses the health care delivery system in each jurisdiction. Each assessment is organized into four parts:

1. an **overview** of the jurisdiction's government, economy, population, and infrastructure;

2. the **organization** of its health care delivery system;

3. the available **health care resources** (financial, health care workforce and technology, supplies, and equipment); and

4. **future health care issues.**

American Samoa

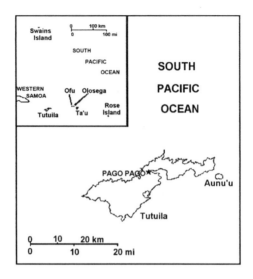

Total Population	58,070
Number of Inhabited Islands and Atolls	7
Access to Major Health Facility (percentage of population requiring more than 1 hour of travel)	50%
Total Health Budget	$21,403,500
Per Capita Health Budget	$369

SOURCE: PIHOA (1997). NOTES: Total population is the official estimate from the 1990 U.S. census; the health care budget is from Fiscal Year 1995.

OVERVIEW

American Samoa is the only U.S. territory south of the equator. U.S. involvement with the islands began more than 120 years ago when U.S. businesses began using the port at Pago Pago. The United States officially annexed the islands in 1900 and placed the U.S. Navy in charge of their administration. In 1951 control was turned over to the U.S. Department of the Interior (DOI). Today, although many aspects of Samoan life have been westernized, the traditional culture—Fa'a Samoa—with its emphasis on extended families headed by a *matai*, or chief, and communal life remains strongly rooted.

The health care system is almost entirely administered and subsidized by the government. On the committee's site visit, the committee heard many expressions of dissatisfaction and frustration with the current state of the health care system from practically every person interviewed—from physicians and nurses to politicians to people waiting at a bus stop. Complaints ranged from concerns about the competence of health care providers and the lack of supplies and equipment to the high cost of sending patients off-island for care.

Government

American Samoans elect a governor and a bicameral legislature (the Fono). They also send a nonvoting delegate to represent their concerns in the U.S. House of Representatives.

Economy

The American Samoan economy is heavily dependent on two industries: government and tuna canneries. DOI estimates that 93 percent of the American Samoan economy is based directly or indirectly on U.S. federal expenditures and the canning industry (DOI, 1996b). In 1993, the American Samoa Government (ASG) employed about 32 percent of the workforce and the canneries employed almost 30 percent (Bank of Hawaii, 1997).[1] The remaining economic activity is based on tourism and small-scale service businesses.

Population

American Samoa's population is increasing rapidly. With an annual growth rate of 3.7 percent, it has one of the highest growth rates in the Pacific (PIHOA, 1997). This growth is fueled by increased immigration and high birth rates. The majority of immigration is from the neighboring islands of Western Samoa and Tonga; additional immigrants have come from the Philippines and other Asian countries. Despite this, traditionally, a considerable amount of emigration has also occurred. Seventy thousand American Samoans now live in the United States (primarily in Hawaii and California); this means that more American Samoans live in the United States than in American Samoa itself (Bank of Hawaii, 1997). The native Samoan population is ethnically Polynesian (rather than Micronesian, as is the case in the other U.S.-Associated Pacific Basin

---

[1]The majority of cannery workers are resident aliens, not American Samoans.

jurisdictions) and represents 90 percent of the island's total population (PIHOA, 1997).

Infrastructure

In general, water, waste disposal, and power systems in American Samoa have improved significantly in the last decade. For example, in 1995, 74 percent of housing units were connected to a public water system, up from 63 percent in 1990. The quality of the water is considered good. Even with these improvements, however, only a little more than half of households have complete indoor plumbing. Electrical power, once highly unpredictable and poorly managed, is provided by new, fuel-efficient electrical plants. Unfortunately, the operations of the privatized American Samoa Power Authority—which is responsible for many of the improvements—are currently threatened by ASG's inability to pay its obligatory subsidies (DOI, 1996b).

Roads, telephones, and distance communications systems (which are responsibilities of ASG) are somewhat in a state of disrepair. According to DOI (1996b), roadways are in poor condition as a result of hurricanes, poor maintenance, and heavy traffic loads. Telephones reach only about 68 percent of households. With expensive long-distance rates, the Internet is currently limited to a privileged few and the community college that uses PEACESAT (although this may change rapidly as a result of new U.S. federal telecommunication regulations).

HEALTH CARE DELIVERY SYSTEM ORGANIZATION

Administration

All health care services are administered through the Department of Health, which is headed by a director of health. The director, who reports directly to the governor, is responsible for overseeing the hospital, public health and dental services, and health planning. Almost all health services are actually provided at the LBJ Tropical Medical Center in Pago Pago, the island's main health care facility and hospital.

The administration of the hospital has changed within recent years. Several years ago, the former governor created a quasi-independent Hospital Authority to oversee the hospital's management and finances. A similar authority had been created for the public utilities a few years earlier with marked improvements in service; it was hoped the Hospital Authority would achieve similar results. The Hospital Authority proved an unpopular, if misunderstood, idea with residents. It was eliminated in January 1997 by the newly elected governor who had run on a platform of getting rid of it. He reinstituted the current administrative structure of having a director of health in charge of all health care services.

Since 1993, an independent consulting team from Mercy International Health Services has also been involved with hospital administration. Originally brought in to co-manage the hospital, their involvement and the scope of the work have changed considerably in recent years. Mercy now acts in a training and advisory role; one of its most important assignments is to find ways to increase reimbursement from Medicare.

Health Care Facilities

The only hospital in American Samoa, the LBJ Tropical Medical Center, is located on the island of Tutuila in Pago Pago. The structure was built in 1968 and has been repeatedly cited by Health Care Financing Administration (HCFA) licensing officials as having major safety problems that could result in life-threatening situations. These problems include serious fire code violations and lack of routine or preventive maintenance. Although plans to correct these deficiencies have been developed, the majority have not been implemented. The situation is serious enough that HCFA has threatened to decertify (and hence stop payments to) the facility on numerous occasions.[2] Some improvements have been made. For example, the Emergency Department was recently updated and remodeled. For the most part, however, over the course of several years few earnest attempts to address the concerns of the HCFA surveyors appear to have been made. Additionally, DOI is currently withholding $2 million dollars in capital improvement funds earmarked for the hospital until an independent authority is established to manage the hospital.

Some community health centers, or dispensaries, exist in outlying villages of American Samoa, although most care is still provided through the hospital (which is relatively accessible by bus and car to most people on the main island). It is unclear exactly how many centers are currently operating; it could be as many as eight. At the time of the committee's site visit, several were reported to have been under construction after being damaged during a hurricane 2 years previously. Immunizations and well-baby care are being provided at these satellite health care centers.

A few private clinics exist. These are staffed by doctors who also work for the government. American Samoa also has three private drug stores and these are owned by the hospital's pharmacist. These arrangements are viewed by many people as presenting potential conflicts of interest.

Health-Related Community Organizations

The American Red Cross is active on the island, helps prepare for emergencies, and provides disaster assistance. Many church groups also provide health-related services such as counseling.

---

[2] As this report was being written, HCFA had begun the process of decertification.

Off-Island Care

Off-island tertiary care referrals consume almost 30 percent of the total health budget and serve less than 1 percent of the total population (PIHOA, 1997). Most patients are referred to Honolulu (private hospitals as well as Tripler Army Medical Center). A patient coordinator in Hawaii helps to coordinate a referred patient's care and to attend to the logistics of lodging for escorts. Consideration is being given to referring more people to New Zealand and Australia as a cost-saving measure.

A Medical Referral Committee (MRC) is supposed to review all cases for off-island referral. The governor must also approve off-island tertiary care referrals. The referrals must be medically necessary and must be for services unavailable on-island. It is reported that the current director of health, however, has stopped referring his patients to MRC and is deciding himself if patients can go off-island.

Many problems have been linked to off-island care. One study found that in 1991 a conservatively estimated 16 percent of all medical referrals were inappropriate with, for example, referrals being granted to family members and friends as political favors (Larin et al., 1994). An increase in the number and cost of off-island referrals was pointed to as one of the major reasons for the government going deeply into debt in the early 1990s (GAO, 1992).

HEALTH CARE RESOURCES

Financial Resources

ASG—and hence the Department of Health—faces grave financial problems. It is several million dollars in arrears with many suppliers and organizations. As of March 1997, it owed Tripler Army Medical Center $1.5 million (P. Barcia, personal communication, March 20, 1997). The U.S. Department of Health and Human Services' discount pharmaceutical supplier, Perry Point, has refused to accept any more requests from American Samoa until it pays its current debts. Most vendors require cash in advance for any new purchase.

The Department of Health annually receives about $2.4 million from Medicaid and $2 million from Medicare, which represents approximately one-fifth of the health budget. Patients are required to make copayments of about $2 for each outpatient visit and $7 a day to stay as an inpatient in the hospital. Collection of fees is rarely pursued. Government employees and cannery workers typically have health insurance and make use of the government health care facilities. Government employees can opt for government health insurance coverage. Virtually no private insurance market exists at this time.

Workforce[3]

Physicians

Four M.D.s and nine medical officers work in American Samoa. Most of the physicians are U.S.-trained expatriates working on short-term contracts (usually for 2 years). The pay scale for health care workers is reported to be one of the highest in the South Pacific region, although many consider it too low to consistently attract high-quality personnel. Additionally, individuals with expertise in some vital areas are lacking. For example, no one on-island is able to surgically repair a blocked shunt for dialysis patients, who must then be sent off-island for this relatively straightforward procedure. Administrative delays are often encountered in the hiring of health care professionals. Doctors from the U.S. Department of Veterans Affairs are sent to American Samoa on a quarterly basis, although they typically conduct eligibility assessments rather than provide medical treatment (GAO, 1993).

Nine medical officers trained at the Pacific Basin Medical Officers Training Program (PBMOTP) are working in American Samoa (two are currently on educational leave at the Fiji School of Medicine). All of them work in the hospital, with occasional duty at the dispensaries. In the future, one or two of the medical officers may be assigned to work full-time for the Public Health Department. Individuals have expressed concerns that the medical officers did not receive enough training in hospital care. In part, this training was to have been provided during the medical officers' required 2-year internship after graduation. The necessary training and supervision for the medical officers within the hospital in American Samoa, however, appear not to have been provided. The lack of appropriate on-island training is one reason given for American Samoa's decision to send its medical officers to the Fiji School of Medicine for further training. Although all PBMOTP graduates can go on to postgraduate training immediately after completing their internship, American Samoan medical officers are being required to repeat 1 or 2 years of undergraduate coursework to earn an M.B.B.S. (the British equivalent of the M.D.). At that point, they can go home to practice or go on to further formal postgraduate education.

Nurses

There are 31 registered nurses, 97 licensed practical nurses, 23 graduate nurses, and 12 nurse's aides currently working in American Samoa. Licensed practical nurses and registered nurses must meet U.S. national certification standards. Many nurses are reported to be unmotivated to take the certification test, particularly because they will still receive pay increases and benefits regardless of whether they have passed the test. Nonetheless, several remedial

---

[3] Workforce numbers are from the 1997 PIHOA Data Matrix (PIHOA, 1997).

programs have been established to help prepare nurses who are planning on taking the national examinations. In 1996, to counter the shortage of practical nurses, the hospital held a successful 15-month training program designed to provide hands-on clinical experience.

The American Samoa Community College (ASCC) has a career ladder nursing program, offering certificates for practical nurses and an associate degree for registered nurses. It also offers an emergency medical technician certificate course through the hospital. This year the nursing program experienced problems recruiting and retaining instructors, primarily as a result of budget cuts at ASCC. Retention of nursing students is also problematic; only 5 nursing students from an original class of 20 students were likely to graduate in 1997. The reasons cited for this poor student retention rate include students' problem with basic English skills and the limited amount of actual clinical experience. Consideration is being given to revitalizing the ASCC program by having it more closely linked to the hospital. Basic academic courses would be provided at ASCC and clinical work would be done in the hospital and dispensaries.

Other Health Care Personnel

Only two of the eight dentists in American Samoa are Samoan, the other six are expatriates on contract (PIHOA, 1997). This may change because nine Samoan students are currently studying at the Fiji School of Dentistry and appear likely to return to American Samoa to practice dentistry. Traditional health practitioners, *fao faos*, are still regularly used by many Samoans. The *fao faos* use herbal remedies and vigorous massage to help their clients.

Workforce Quality Assurance

As mentioned earlier, HCFA is responsible for monitoring quality in the hospital, and its surveyors have repeatedly determined that quality and safety are seriously lacking. Although a few attempts to establish a quality assurance program in the hospital have been made, no comprehensive program is in place. No continuing education is required for any health care provider. A Health Services Regulatory Board was created to license all health professionals (except nurses) and to ensure that certain minimum educational and professional criteria are met. However, the board has not met in many months and is considered to be defunct. Several of the PBMOTP graduates do meet on their own to stay abreast of the literature and discuss research. Nurses have advocated for legislation to implement a Nurse Practice Act, but this has not been passed by the Fono.

Technology, Supplies, and Equipment

Shortages of even the most ordinary supplies such as X-ray film and pain medication are a chronic and recurring situation at the hospital. Shortages of supplies have forced the dialysis unit to be closed for up to a week, imperiling patients' lives. Staff have sometimes been creative in improvising ways to deal with shortages. For example, when supplies of oxygen ran out, staff arranged for a federal surplus oxygen generator to be installed. Although the gauge showing the actual quality of the oxygen is broken, the new equipment has the potential to save thousands of dollars each year.

A mammogram machine was donated to the hospital several years ago, but it is not being used because it is broken and no one is trained to operate it. A computed tomography (CT) scanner has been requested as a way of reducing the costs related to sending people off-island for diagnosis. There would need to be a person trained in the machine's operation and maintenance as well as the clinical skills needed to interpret the images. The LBJ Tropical Medical Center has several dialysis machines and hopes to add more machines in the near future to meet increased demands.

PEACESAT provides an opportunity for limited telemedicine applications, but to date this has been used mostly for in-service training of physicians. During the committee's site visit, however, the equipment was broken and a scheduled session was canceled indefinitely.

FUTURE HEALTH CARE ISSUES

The most pressing issue confronting the American Samoan health care system is getting its financial house back in order. The hospital's physical plant must be brought up to minimum safety requirements. Vendors need to be paid so that shortages of vital equipment and supplies no longer imperil patients' lives. Residents of American Samoa are not receiving the quality health care they so strongly desire.

Commonwealth of the Northern Mariana Islands

[Map showing islands: Farallon de Pajaros, Maug Islands, Asuncion Island, Agrihan, Pagan, Guguan, Sarigan, Anatahan, Farallon de Medinilla, Saipan, Tinian, Rota, within the Philippine Sea and North Pacific Ocean.]

Total Population	58,846
Number of Inhabited Islands and Atolls	3
Access to Major Health Facility (percentage of population requiring more than 1 hour of travel)	90%
Total Health Budget	$36,161,007
Per Capita Health Budget	$614

SOURCE: PIHOA, 1997. NOTES: Total population is the official estimate from the CNMI 1995 mid-decade census; health care budget is from Fiscal Year 1996.

OVERVIEW

The Commonwealth of the Northern Mariana Islands (CNMI) has witnessed tremendous economic and social changes since 1978, when it began the process of becoming an official commonwealth of the United States. The tourism, garment, and construction industries expanded rapidly, creating a labor shortage. Tens of thousands of workers from neighboring countries moved to CNMI to get jobs. Although incomes soared, the rapid and largely unplanned economic development and resulting influx of foreigners have put stress on health care and other basic services. Development has also been cited as a factor in the breakdown in traditional family arrangements and an increase in the use of alcohol and drugs.

Government

Although they do not vote in U.S. elections, CNMI residents are U.S. citizens and elect their own governor, lieutenant governor, and a legislature with nine senators and 15 representatives. Although they have a representative in Washington, D.C., that person is a not an official congressional delegate like those from American Samoa and Guam. Because CNMI is a U.S. commonwealth, its government has control over many policies such as immigration, tax, and labor.

Economy

CNMI's economy is based largely on tourism and the garment industry. In 1995, almost 655,000 tourists—primarily from Japan—visited CNMI and spent roughly $522 million (DOI, 1996b). That same year, the garment industry exported over $419 million in goods, primarily to the United States. Concerns about possible labor abuses and poor working conditions in both the public and private sectors have brought increased attention from the United States and several foreign governments. Although once used by the Central Intelligence Agency as a training base, CNMI currently has no U.S. military bases and hence no income from the U.S. military, unlike other U.S.-Associated Pacific Basin jurisdictions.

Population

Within the past decade and a half, the CNMI population has exploded—from 16,890 in 1980 to 58,846 by 1995 (U.S. Bureau of the Census, 1997b; PIHOA, 1997). Almost all of the growth results from immigration—workers for the booming tourism and garment industries—primarily from the Philippines and China. In fact, today for every native CNMI resident there are approximately two non-resident aliens. Of the native population that does exist, 75 percent is Chamorro and the rest is Carolinian.

Infrastructure

Water, wastewater, and sewage systems operate at the limit of their capacity on Saipan, the main island. Although millions of dollars in improvements have been made to the public water system since 1990, both the World Health Organization (WHO) and the Health Care Financing Administration (HCFA) have deemed the water supply unsafe (HRSA, 1996). With no rivers and few springs, rainfall provides most of the fresh water. In addition, excessive well-drilling and cracked pipes in the distribution system have produced salty, unhealthy water in the groundwater lens (DOI, 1996b).

Heavy rainfall often floods the wastewater collection system on Saipan, causing waste to rise and dissipate with the rain. In addition, Saipan's only waste disposal site, the Puerto Rico dump, has frequent fires that release toxic gases, threatening residents in surrounding areas. Plans are underway, however, to close the dump and to open a new landfill, for which $16 million in Covenant and local funding has been earmarked (DOI, 1996b).

Transportation on the island of Saipan is also troublesome. Many of the 362 kilometers (225 miles) of roads throughout CNMI were paved during World War II and are now being rebuilt. Primary highways are increasingly overburdened with heavy traffic. Traffic lights have been added, and plans to reconstruct and pave more roadways are underway. Even so, the major road on Saipan, Beach Road, is often inundated with traffic jams, and motor vehicle accidents are one of the leading causes of death (DOI, 1996b).

HEALTH CARE DELIVERY SYSTEM ORGANIZATION

Administration

Health care services are administered by the Department of Public Health. The department operates the Commonwealth Health Center (CHC), the hospital and main outpatient center, and administers all public health, mental health, and related social service programs. In recent years it has actively begun to privatize some of its services, such as outpatient pharmacy, linen and laundry, yard and grounds maintenance, security, and nurse recruitment. The private sector, in general, is expanding rapidly in CNMI. Several private health clinics operate on the islands, some in conjunction with private health maintenance and insurance organizations.

Health Care Facilities

CHC serves as the main hospital for CNMI. Built in 1986, the 76-bed facility provides both acute inpatient and outpatient services. CHC has 13 hemodialysis machines and a computed tomography (CT) scanner. As a recipient of Medicare funds, CHC is subject to the licensing and certification of HCFA. The facility has fared well on past surveys and has followed through in implementing its plans of corrections. CHC is also the only health care facility in the region to be fully accredited by HCFA.

APPENDIX D 107

Government-run health centers have been established on the islands of Rota and Tinian, and a new clinic operates in the southern village of San Antonio on Saipan. The emphasis in these health centers and clinic is to provide preventive services such as immunizations, prenatal care, and other primary health care services. Three private health clinics and one dental clinic have also been opened in recent years.

Health-Related Community Organizations

The CHC Volunteers is a volunteer group dedicated to raising funds for CHC. Money is raised through the hospital gift shop, raffles, and other special fundraising activities. In the past 10 years, CHC Volunteers has raised more than $750,000. The money has been used not only to buy needed equipment in the hospital, but also to sponsor a television series on health education and other community health information efforts. Other community organizations include youth drug prevention groups, the CNMI National Food and Nutrition Advisory Council, and a variety of school health education programs in several schools in CNMI.

Off-Island Care

CNMI recently enacted legislation that requires decisions about off-island tertiary care referrals to be made by a Medical Referral Committee (MRC) comprising six physicians. This has helped to depoliticize the process; on the site visit the committee was told that a senator's mother's request to have a gallstone operation off-island had been turned down by the MRC which felt that the operation could be appropriately handled on-island (it was, and with a good result). Typical reasons for off-island referrals include receiving services not available on-island, such as open-heart surgery or chemotherapy. People who do go off-island for medical treatment at government expense must now contribute toward the cost on a sliding fee scale and are subject to a $50,000 cap for expenses. The CNMI government budgeted $7 million in 1996 for the off-island referral program and included the incentive that if any money for the referral program remains at the end of the year it would be turned over directly to the health department (rather than the government's general fund).

Health Care Resources

Financial Resources

The total health care budget in 1996 was $43 million, which represents 15 percent of the CNMI government's total budget. This includes funds from local taxes ($28 million), a CNMI government legislative appropriation ($10 million),

Medicare and Medicaid ($1.7 million), U.S. federal grants ($2.4 million), and WHO ($100,000) (I. Abraham, personal communi-cation, August 21, 1997).

By law, health services must be provided to all regardless of their ability to pay. However, private health insurance is available through a number of companies and health maintenance organizations. It is estimated that roughly 60 to 70 percent of the population has some form of private health coverage. Employers of foreign workers are required to provide insurance to their employees (PIHOA, 1997).

Compact Impact

Under the provisions of the Compacts of Free Association, residents of the freely associated states may enter CNMI of their own accord. Many have done so in search of economic opportunities. In 1990, an estimated 3,327 people born in the freely associated states resided in CNMI, a little over half of whom immigrated after the Compacts were signed in 1986 (Levin, 1996). Although many are employed and have health insurance or sufficient income to cover the cost of their own health care, the CNMI Department of Public Health estimated that in 1996 it provided health care costing $1,480,000 to citizens of the freely associated states. This represents 11 percent of all the encounters at government-operated health facilities (CNMI Department of Public Health, 1997).[4]

Workforce

Physicians

The physician workforce in the CNMI is largely expatriate, with more than 90 percent of doctors coming from outside the region. Recruitment on the international market has been difficult in the past, but the pay scale and the benefits now being offered (government-sponsored housing, transportation and moving allowances, etc.) are believed to be competitive. Government-employed physicians are covered for malpractice, but private physicians are not. The CNMI government chose not to participate in the Pacific Basin Medical Officers Training Program (PBMOTP) because of concerns about meeting HCFA licensing requirements and local laws of the CNMI Medical Profession Licensing Board. An informal program to encourage local students interested in becoming doctors and practicing medicine in CNMI has recently been undertaken. Students study in the United States, are offered summer jobs in CHC to gain experience, and are given a mentor in CNMI who stays in contact with the student while he or she is studying on the U.S. mainland.

---

[4]For a more complete discussion of "Compact Impact" and references to relevant literature, the reader is referred to Levin, 1996.

Nurses

Recruitment and retention of nurses are problematic. Currently, the overwhelming majority of the nurses and assistants are foreign contract workers (primarily from the Philippines) hired through employment agencies. Many of these workers only stay long enough to complete a 2-year contract, get training, and pass the U.S. licensing examination (NCLEX), before moving to the United States or elsewhere. The Northern Marianas College has an accredited 2-year nurse degree program. Nursing students receive clinical experience at the CHC. The CHC itself provides training for nurses challenging the NCLEX, as well as a variety of continuing education courses. Nonetheless, with a limited pool of nurses being trained on-island, nurses continue to be actively recruited from Palau, the Philippines, and other neighboring islands.

Workforce Quality Assurance

Both physicians and nurses must meet certain quality standards to be licensed by the CNMI Medical Profession Licensing Board. As mentioned earlier, HCFA also accredits the hospital.

Technology, Supplies, and Equipment

CNMI will soon have fiberoptic capability for telecommunications. CHC is ready to hook up with Hawaiian and U.S. mainland hospitals for clinical telemedicine consultations as soon as the fiberoptic connection is completed (I. Abraham, personal communication, April 18, 1997). The Northern Marianas College already uses a microwave system to provide interactive distance education courses from Saipan to students in Tinian.

FUTURE HEALTH CARE ISSUES

CNMI plans to continue finding innovative ways to provide health care for its rapidly growing population. It also plans on continuing to promote and develop the private health care sector. At the same time, it remains committed to providing quality health care to everyone, regardless of the ability to pay a commitment backed by the CNMI government. Finally, in the words of Secretary Abraham, "The most important vision we have for the future is a community whose members understand they must take responsibility for their health and health care" (Abraham, 1997).

Federated States of Micronesia

```
PHILIPPINE                    NORTH              MARSHALL
  SEA          GUAM           PACIFIC             ISLANDS
                              OCEAN

               C A R O L I N E   I S L A N D S
     Ulithi
   Yap    Fais  Gaferut
   Islands        Farauep  West Fayu    Fayu  Murilo
            Sorol          Pikelot  Namonuito  Nomwin  Minto
  Ngulu     Woleai        Satawai  Pulap   Chuuk   Pakin  Palikir  Mokil
            Eauripik  Ifaluk          Lagoon  Oroluk  Pohnpei
                              Puluwat  Namoluk                Pingelap  Kosrae
      YAP                          Satawan
                                        Nukuoro
                       CHUUK                                 KOSRAE
                                            POHNPEI
   0    200   400 km
   0       200     400 mi         Kapingamarangi
```

Total Population	105,506
Number of Inhabited Islands and Atolls	62
Access to Major Health Facility (percentage of population requiring more than 1 hour of travel)	(See below under descriptions of individual states)
Total Health Budget	$13,962,807
• National Health Budget	1,177,441
• State Health Budgets	12,785,366
Per Capita Health Budget	$132

SOURCE: PIHOA (1997). NOTES: Total population is the official estimate from the FSM 1994 National Census; the health care budget is from Fiscal Year 1994.

OVERVIEW

An independent country since 1986, the Federated States of Micronesia (FSM) consists of four states: Chuuk, Kosrae, Pohnpei, and Yap. With the exception of Kosrae state, which consists of only one island, all the other states have many islands scattered across vast stretches of the Pacific Ocean. Almost all health care services are provided by the federal and state governments of FSM, although the quality of the health care system varies markedly from one state to another. The FSM economy—including its health sector—is extremely dependent on U.S. aid and, as funding from the Compact of Free Association winds down, uncertainty prevails across many aspects of island life.

Government

The government structure is very similar to that of the United States with federal, state, and local governments. Both the federal and state governments of FSM have separate executive, legislative, and judicial branches. Traditional leaders continue to play key roles in the decisionmaking process.

Economy

The FSM economy depends heavily on its public sector. The government, largely supported by U.S. funds, provides most of the highest paying employment opportunities, especially in the more urban areas. Roughly 25 percent of the population works for the national or state governments of FSM (FSM, 1991). A more traditional subsistence economy based on farming and fishing dominates the more remote outer islands and villages.

Serious planning for the country's economic future has begun in earnest as the Compacts wind down. The Asian Development Bank is working with the federal and state governments; each state has held a summit to develop economic plans and priorities, including plans and priorities for the health sector. Emphasis is being placed on developing FSM's fishing, tourism, and agriculture industries.

In the short term, however, the federal and state governments have begun to change their personnel policies in anticipation of further decreases in funds from the United States. For instance, many government employees, including health care workers, are now on a 4-day work week, and salaries for government workers have been reduced by 20 percent. In light of the limited economic opportunities, it is assumed that many FSM residents will emigrate to Guam, the Commonwealth of the Northern Mariana Islands, Hawaii, and the U.S. mainland—a trend that has been on the rise since the Compacts first went into effect and presumably will still be allowed when the Compacts are renegotiated (DOI, 1996a).

Population

The native FSM population is primarily Micronesian and includes the major ethnic groups of the individual states: Chuukese; Kosraean; Pohnpeian; Yapese. Overall, the population is very young with 44 percent of the population under 15 years of age. The population growth rate has decreased to 2 percent in recent years. The crude birth and fertility rates are high compared with the rates in more developed countries; on average, women in FSM have 4.7 children (PIHOA, 1997). This rate is much lower than the 8.2 children women averaged in 1973, however (FSM, 1996). The decline in the population growth rate is most directly attributable to high infant mortality rates (46 per 1,000 births) and emigration from FSM (PIHOA, 1997; Hezel et al., 1997). Family planning, and the education and employment of women are other important factors typically associated with such a change (Marshall and Marshall, 1983).

Infrastructure

Considered a private-sector success model, FSM has turned over road maintenance, telecommunications, and power to private companies. Water, sewage, and waste disposal remain the responsibility of the Department of Public Works.

The level of access to sanitation facilities on FSM is extremely low, and this is particularly a problem in the crowded urban centers of Chuuk and Pohnpei. According to the U.S. Department of the Interior, in 1995 only about 34 percent of households in FSM had flush toilets, about 18 percent were connected to a public water system, and 11 percent were connected to a public sewer. Electrical power is available to only about half of the households (or 51 percent), the lowest number among the six jurisdictions of the U.S.-Associated Pacific Basin (DOI, 1996b). (See also the descriptions in the assessments of the individual states given below.)

HEALTH CARE DELIVERY SYSTEM ORGANIZATION

Administration

FSM has a national Department of Health located in the capital, Palikir, which is responsible for overall health planning and technical assistance. Each state has its own state department of health, which provides actual health care services through a central hospital and a variety of primary care delivery sites (dispensaries), which can range from a building with several rooms to a medicine cabinet in a person's home. Typically, these state departments of health administer an inpatient ward, an outpatient department, a dental department, and a public health department. The public health departments conduct all of the current preventive activities (well-baby care, prenatal care, family planning, sexually transmitted disease treatment and follow-up of contacts, tuberculosis and leprosy treatment and follow-up of contacts, and any activities—admittedly sparse—devoted to the follow-up of chronic diseases like hypertension and diabetes). In some jurisdictions, within this public health department is a separate entity that specifically oversees the peripheral clinics or dispensaries.

Only a handful of practitioners are in private practice; the rest work for the federal or state government. Further development of private practices has been hampered by the need to use government facilities for laboratory work or inpatient hospital care. Legislation regarding liability and malpractice is also needed before more privatization can occur.

Health Care Facilities

(See descriptions in the assessments of the individual states given below.)

Health-Related Community Organizations

Each jurisdiction has a number of community organizations that are involved with some aspects of health care. Some churches and women's and youth groups offer such services as supportive counseling and a mechanism for health promotion activities (Hezel et al., 1997). The Red Cross is active in Chuuk.

Off-Island Care

All four states have protocols and medical review committees to review requests for off-island tertiary care referrals. Many states employ staff in the major referral centers (such as Honolulu and Manila) to coordinate care. In the past these referrals have drained significant portions of the total health care budgets to serve a very small percentage of the overall population. For example, in 1992 Chuuk spent $472,000 and overran its budget for such referrals by $30,000 (PIHOA, 1997). Reportedly, the decisions of the medical review committee are often overruled by members of the legislature. In Kosrae, both the number of cases and the total amount spent on off-island referrals has decreased in recent years, going from 34 cases accounting for 12 percent of the total health care budget in 1990 to 20 cases and 10 percent of the total health care budget in 1996 (Kosrae Department of Health Services, 1996).

Nonetheless, on the site visits committee members were told by FSM health officials tht a recent study from an outside reviewer reportedly found that 92 percent of all referrals were medically justified [need to get cite from Dr. Pretrick]. The remaining 8 percent still represents a significant and inappropriate drain of health care dollars away from the jurisdictions, even though the overall trend appears to be a steady albeit slow and difficult movement away from the abuses and cost overruns of the past.

In April 1997, the FSM government revised its off-island medical referral program for government employees. Members of the National Government Employee Health Insurance Plan (NGEHIP), which is the government-subsidized health insurance, must now use accredited providers or hospitals in the Philippines and Hawaii (a private medical group and the Tripler Army Medical Center). Referral patients who choose not to go to these facilities and providers are responsible for paying all their medical bills up front. Reimbursement for such billings is then based on the same regulated fee schedule being followed by all other off-island accredited facilities; the difference is the individual's responsibility. So, for example, if a person needs a procedure that costs $1,000 at Tripler Army Medical Center, but decides that he wants to have it done in San Diego, where the same procedure costs $4,000, he will have to cover his own airfare and pay the $3,000 difference.

HEALTH CARE RESOURCES

Financial

Underscoring its heavy dependence on U.S. aid, the FSM Department of Health estimates that if Compact funding ends as scheduled and no more U.S. aid is provided, funds for health services would decline by as much as 75 percent of the 1996 health budget (FSM, 1996). It is also estimated that FSM will receive little aid from other foreign countries; in 1997, it received $748,822 (or approximately five percent of its total health care budget) in non-U.S. foreign aid for health (Samo, 1997).

FSM laws require that no person be denied health care because of an inability to pay. Although each jurisdiction has established some cost-sharing requirements, the amounts are very low and collection is not vigorously pursued. In Chuuk, for example, the average amount collected from the primary care delivery sites (dispensaries) is only $0.70 a month. The fees that are collected for health care services go to the state's general account.

Government employees can voluntarily decide to participate in NGEHIP. Virtually no private health insurance market exists at present.

Workforce

(See descriptions of health care workforces in the assessments of the individual states given below.)

Quality Assurance

The Micronesian Medical Council provides for the nominal licensure of health professionals in the region, although no one is required to become licensed. Continuing education and training courses for physicians and medical officers are offered through the regionwide Pacific Basin Medical Association, as well as the newly organized medical associations in each FSM state.

Technology, Supplies, and Equipment

Computers are available in the hospitals of FSM, but they are not often used. The cost of access to the Internet used to be prohibitive, but in recent months it has decreased considerably (in some places as little as $20 a month). Several hospitals are equipped with donated Picasso telephones that use a still-image, store-and-retrieve method of telemedicine. However, at $3/minute the cost of transmitting these still images has proven too costly for at least one jurisdiction (Kosrae) (WPHealthnet Infoline, 1997). It is unclear the extent to which the Picasso phones are currently being used. Other forms of telemedicine are virtually nonexistent, although there have been efforts recently to use the PEACESAT satellite for monthly conference calls for information sharing among health professionals throughout the region.

Supply, equipment, and drug shortages are chronic. Sophisticated laboratory equipment is available in the hospitals, but it is often unusable because the necessary reagents and other supplies are unavailable. Some reasons for this include the inefficiency of the procurement process and personnel, unreliable air transportation, reprogramming of funds, and use of high-priced third-party vendors and emergency orders.

The lack of maintenance of equipment is also an area causing grave concern. For example, on the committee's site visit committee members were told that the majority of X-ray machines (i.e., those that are still working) in FSM have not been properly calibrated in years. The quality of the lead lining in some hospitals is unknown. Yet radiology staff in some of the jurisdictions do not routinely wear badges indicating how much radiation they have been exposed to or turn their badges in for analysis.

FUTURE HEALTH CARE ISSUES

The FSM Department of Health identifies the following issues as its priorities over the next 5 years (FSM, 1996):

- **Health care financing**: develop a secure financial base for health services and a capacity for improved cost accounting.
- **Primary health care systems:** move the focus from curative to preventive health services, which emphasize the health needs of women and children, limit population growth, and promote healthy lifestyles.
- **Epidemiological surveillance and data collection:** improve the methods of and capability for measuring and assessing the population's health needs and conditions.
- **Quality assurance:** develop minimum facility and workforce standards and regulations to ensure a basic level of competent and safe care.
- **Environmental, food, and drug safety:** improve all of these to protect the nation's health.
- **Off-island tertiary care referrals and diagnostic services:** develop means of controlling the spiraling costs associated with this type of care.

Chuuk

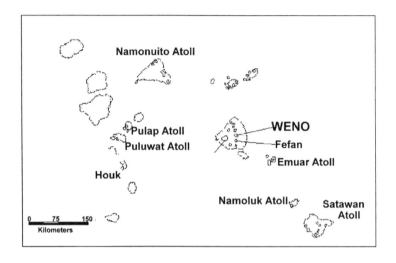

Total Population	53,319
Number of Inhabited Islands and Atolls	40
Access to Major Health Facility (percentage of population requiring more than 1 hour of travel)	62%
Total Health Budget	$4,884,786
Per Capita Health Budget	$92

SOURCE: PIHOA (1997). NOTES: Total population is the official estimate from the 1994 FSM National Census; health care budget is from Fiscal Year 1994.

INFRASTRUCTURE

Like other FSM states, Chuuk has benefited from some of FSM's improvements to island infrastructure and privatization of its utilities. As the most populated and poorest state, however, Chuuk continues to suffer from poor sanitation, water shortages, and unreliable power. Drinking water, for example, is generally accessed through wells to an underground lens, and overdrilling in the past has damaged the water supply (FSM, 1991), and there are concerns about the future availability of underground water (Bank of Hawaii, 1995a). At the same time, on the site visit committee members were told that an estimated 40 percent of the municipal water supply in Weno is lost through water leaks. In addition, Chuuk's geographic location and low-lying atolls make it particularly vulnerable to typhoons and severe weather conditions that can easily damage homes, power, and facilities. Lack of roadways and poor telecommunications make it difficult to expand a potentially lucrative tourist market (Bank of Hawaii, 1995a).

Health Care Facilities

The Chuuk State Hospital, a 125-bed facility built in 1969, is located in Weno. It is in serious disrepair. The wooden structure is also considered to be a fire risk and has a major termite infestation. Fire alarms do not work, and the water pressure throughout the hospital is extremely low. Broken equipment is scattered throughout the facility. Electricity is unreliable throughout the island, although the situation at the hospital has been aided since Queens Hospital in Hawaii donated a backup generator. Not all wards, including the pediatric ward, are hooked up to the backup generator, however; people must resort to using candles for light, increasing the risk of fire. Supply shortages are commonplace and chronic; during the site visit there were no supplies in the laboratory or X-ray departments. In the month before the committee's site visit a ventilator-dependent patient had died because oxygen supplies ran out.

Renovation and expansion of the pediatric ward has started, but is on hold until the Chuukese government can find more funds to compete the work. Once the work is completed, U.S. Department of Interior has agreed to reimburse the expenses. In the meantime, the demolished ward continues to provide a hazardous pass-through to other hospital wards. Although there is no money and no master plan to do so, officials hope to build a new hospital rather than refurbish the current structure.

Although the government lists 67 dispensaries (PIHOA, 1997), according to a recent report, only about half appear to be in operation (Medical Graduate Support Program, 1997). Even those that are reported to be in operation often have no supplies—even of the most basic medications like aspirin—because most supplies come from the Chuuk State Hospital, which itself is almost always out of supplies. Some health assistants resort to paying for supplies themselves to meet the needs of the community. According to the report, 21 dispensaries are run out of the health assistants' homes, a situation described as "the same as none" (Medical Graduate Support Program, 1997, p. 10). Health assistants are paid a salary whether they work or not and are also paid rent if they work out of their homes. Most of Chuuk's dispensaries have been built on private land and must be leased. This causes problems when the government does not pay its bill or when the title to the land is not clear. Remote outer islands have access to radios in government offices to contact the hospital in the event of an emergency. Residents on islands closer in can transport people to the hospital via boat.

Workforce[5]

Physicians

Chuuk has five M.D.s (all expatriates from the Philippines and Burma) stationed at the hospital. They also have 19 medical officers (14 who trained at the Pacific Basin Medical Officers Training Program [PBMOTP] who also work in the hospital. One PBMOTP graduate is at Fiji doing postgraduate work in obstetrics.

Dentists

The dental workforce comprises five expatriate dentists, one indigenous dental officer nearing retirement, one dental therapist, three dental nurses, and 11 dental aides.

Mid-Level Practitioners

Eighty-four health assistants (including five medexes) staff the primary care delivery sites (dispensaries). Most have less than a high school education and have not received ongoing training and education. Supervision of the health assistants appears to be minimal.

Nurses

With 55 registered nurses and 99 licensed practical nurses, Chuukese officials believe there is a nursing shortage. The inability to recruit trained nurses and the loss of Chuukese nurses to other islands where the pay is better is a problem. Poor supervision of nurses and high absenteeism also contribute to the perceived shortage of nurses.

---

[5]Workforce data are from the 1997 PIHOA Data Matrix (PIHOA, 1997).

Kosrae

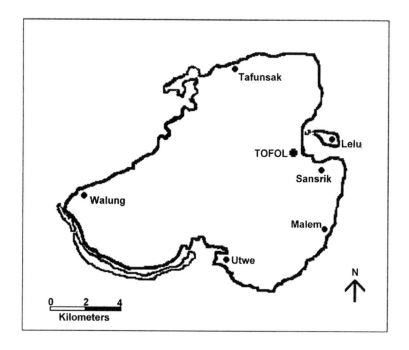

Total Population	7,317
Number of Inhabited Islands and Atolls	1
Access to Major Health Facility (percentage of population requiring more than 1 hour of travel)	100%
Total Health Budget	$1,104,444
Per Capita Health Budget	$151

SOURCE: PIHOA (1997). NOTES: Total population is the official estimate from the 1994 FSM National Census; health budget is from Fiscal Year 1994.

INFRASTRUCTURE

As the least populated and the only single-island state in FSM, Kosrae enjoys comparatively stronger infrastructure systems. Fresh water from perennial streams provides much of the drinking water (Hezel et al., 1997). Most households have access to electricity, and new roads that circumnavigate the island, improvements to ports, and expansion of the airport in the late 1980s have helped Kosrae to attract some foreign trade and small-scale tourism (FSM, 1991).

Health Care Facilities

A 40-bed facility built in the mid-1970s is located in the main village of Tofol. It is in the process of being renovated. Two health centers in outer villages have recently started operations, and two other such health centers are planned. The health centers are designed to provide preventive services and basic medical care and are staffed by medical officers. The responsibility for managing the centers is planned to be turned over gradually from the state government to the villages. At this time, however, only very limited services are being provided at the health centers, and these are provided only on an infrequent basis.

Workforce

Almost all of the health care workforce (96 percent) in Kosrae is indigenous (Kosrae Department of Health Services, 1996).[6] Recently, the work week for all state employees was reduced to 3 and one-half days.

Physicians

The only M.D. on Kosrae is a surgeon from the Philippines. One native Kosrean with an M.B.B.S. is also practicing. Seven graduates of PBMOTP are now working in Kosrae, and two of them have received postgraduate education at the Fiji School of Medicine (one in children's health and one in anesthesiology).

Dentists

Only one dentist serves the needs of the island. He is aided by one dental therapist, three dental nurses, and one dental aide.

Mid-Level Practitioners

Currently, the only mid-level health care workers are three health assistants.

---

[6]Workforce data are from the 1997 PIHOA Data Matrix (PIHOA, 1997).

APPENDIX D

Nurses

All the nurses are Kosrean and are graduates of the College of the Marshall Islands School of Nursing. There are 34 graduate nurses, 3 licensed practical nurses, and 1 registered nurse.

Pohnpei

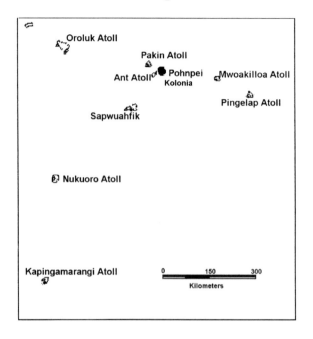

Total Population	33,692
Number of Inhabited Islands and Atolls	6
Access to Major Health Facility (percentage of population requiring more than 1 hour of travel)	77%
Total Health Budget	$4,808,691
Per Capita Health Budget	$143

SOURCE: PIHOA (1997). NOTES: Total population is the official estimate from the 1994 FSM National Census; health budget is for Fiscal Year 1994.

INFRASTRUCTURE

Pohnpei, as with Chuuk, contends with public health concerns like infectious, diarrheal, and skin diseases that occur in crowded conditions, lack of potable water, and poor sanitation. Outside the main center, most of Pohnpei's households still do not have access to public water or sewage systems. One recent study that sampled households in six of Pohnpei Island's villages reported that a majority of homes did not have access to waste disposal systems; most contained open-air pit latrines. Sixty-four percent of homes gathered their water

APPENDIX D   123

from rivers, others through various catchment systems (Yaingeluo, 1997). Sewage disposal systems are another problem: in Kolonia, Pohnpei's main landfill is leaking into an adjacent lagoon near the Kolonia airport (Hezel et al., 1997). One notable improvement in Pohnpei has been its telecommunications system, noted as one of the best in the Pacific region (DOI, 1996b).

Health Care Facilities

This 91-bed facility was built in the mid-1980s and, although a group of volunteers has painted it recently, it suffers greatly from a lack of routine facility maintenance.

Reorganized in 1996, the Division of Primary Health Care operates a network of six primary care delivery sites (dispensaries) on Pohnpei and five on outer islands. One of these, a Section 330 community health center, operates at two locations on Pohnpei. All these sites receive regular visits from the medical officers and are staffed by health assistants and nurses. Although relatively new and extremely susceptible to budget cuts, the move towards greater decentralization reverses the trend in the 1980s to do away with dispensaries and move medical staff to the main island and hospital.

Workforce[7]

Physicians

Six expatriate M.D.s, one indigenous M.D., and two indigenous M.B.B.S.s are currently in practice. Fourteen medical officers are also practicing, thirteen of whom are PBMOTP graduates. One PBMOTP medical officer is completing postgraduate training in anesthesiology at the Fiji School of Medicine. One private practitioner works in Pohnpei, as well as one ophthamologist.

Dentists

Four expatriate dentists (from Burma and the Philippines) and two indigenous dental officers work on Pohnpei. They are assisted by ten dental nurses and nine dental aides.

Mid-Level Practitioners

Four medexes and 15 health assistants round out the health care workforce.

---

[7]Workforce data are from the 1997 PIHOA Data Matrix (PIHOA, 1997).

Nurses

There are 63 registered nurses and 49 licensed practical nurses, almost all of whom are indigenous. Pre-nursing coursework is offered through the College of Micronesia.

Yap

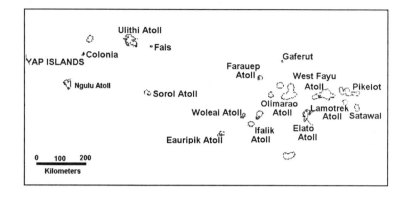

Total Population	11,178
Number of Inhabited Islands and Atolls	15
Access to Major Health Facility (percentage of population requiring more than 1 hour of travel)	65%
Total Health Budget	$1,987,445
Per Capita Health Budget	$178

SOURCE: PIHOA (1997). NOTES: Total population is the official estimate from the 1994 FSM National Census; health budget is for Fiscal Year 1994.

INFRASTRUCTURE

Most people within the capital center of Colonia have access to a public water system. However, while on the site visit committee members were encouraged not to drink the tap water. A recent report noted that the Yap landfill, which sits atop a hill, lies in close proximity to the only water reservoir (Hezel et al., 1997).

The telecommunications infrastructure has improved on the main island; some government workers were even connected to the Internet and were using it to track an approaching typhoon while the committee visited the island. It was also evident, however, that access to telephones was not common, and long-distance connections often failed or were of poor quality. Communication with outer islands is done through solar-powered shortwave radio. A PEACESAT station is located at the Department of Education (PIHOA, 1997).

Health Care Facilities

Yap State Hospital is a 43-bed hospital facility located in Colonia. The Primary Care Program in the Public Health Division oversees a network of 30 primary care delivery sites (dispensaries), considered at one time by many observers to be a model of primary health care delivery. Community involvement with some of the dispensaries is quite high—from villagers assisting with building and maintaining the facilities to a community board providing guidance on the day-to-day operations. All dispensaries had been equipped with solar-powered shortwave radios so that they can communicate with a primary care team located at the hospital. The primary care team also makes occasional field trips to the outer islands.

However, a recent report states that only 3 or 4 of an original 13 primary care sites are still in operation in Yap proper (in Makiy, Thol, Rumung, and possibly Gilman) (Medical Graduate Support Program, 1997). The reasons for this decline include improved access to the hospital, availability of other dispensaries nearby, decrease in funding from international sources, lack of community support for them as a priority, and a lack of trained personnel and medical support. The other 17 sites, located on the outer islands, appear to still be in operation—although the number of field trips with the primary care team has decreased. (J. Gilmatam, personal communication, September 16, 1997).

Workforce[8]

Physicians

Yap's physician workforce consists of 17 physicians: 1 expatriate M.D. (a National Health Corps Service doctor, board-certified in family practice), 4 indigenous M.B.B.S.s and 1 expatriate M.B.B.S., and 11 medical officers, 7 of whom are PBMOTP graduates.

Dentists

Yap has only one dental officer and he is about to retire; he is aided by 10 dental nurses.

Nurses

Although only one registered nurse is currently practicing in Yap, there are 12 graduate nurses, 12 licensed practical nurses, and 3 nurses aides. The University of Guam hopes to extend its distance education courses for nurses (currently being offered in Palau) to Yap within the next year.

---

[8]Workforce data are from the 1997 PIHOA Data Matrix (PIHOA, 1997).

Health Assistants

Yap has approximately 30 health assistants, most of whom staff the outer-island primary health centers (dispensaries). Some of these health assistants may have been trained as medexes (Medical Graduate Support Program, 1997).

Guam (Guahan)

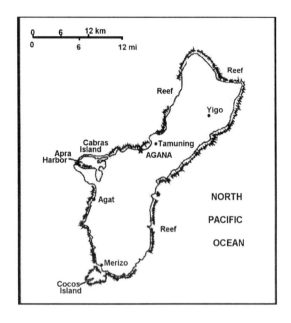

Total Population	155,225
Number of Inhabited Islands and Atolls	1
Access to Major Health Facility (percentage of population requiring more than 1 hour of travel)	90%
Total Health Budget	$81,000,000
Per Capita Health Budget	$510

SOURCE: PIHOA (1997). NOTES: Total population is the official estimate from the 1990 U.S. census; health budget is for Fiscal Year 1998.

OVERVIEW

The U.S. Territory of Guam is the largest and southernmost of the Mariana Islands. It straddles the western Pacific as at once a leader and an island caught in transition. By comparison with the other U.S.-Associated Pacific Basin jurisdictions, Guam ranks above the other island states in overall wealth and health. However, the island still lags behind U.S. mainland states, Alaska, and Hawaii in many of these indicators. This duality presents one of the greatest challenges to Guam as it strives to improve the health and well-being of its people.

Guam's relationship with the United States dates back nearly 100 years. Along with the islands of Puerto Rico, Cuba, and the Philippines, Guam was acquired as a U.S. possession following the Spanish-American War in 1898. Until 1949, and with the exception of 30 months of Japanese occupation during World War II, Guam remained under U.S. military administration. In 1949 administrative control moved to the U.S. Department of the Interior. In 1950 the United States and Guam passed the Organic Act that established a local government on Guam. However, the U.S. military has maintained a strong presence on the island, retaining about one-third of Guam's land for its use.

Government

Much like its Pacific neighbors, Guam's political and social structures are rooted in years of U.S. administration. Guam is an organized, unincorporated territory of the United States. The executive branch consists of a governor and a lieutenant governor. The legislative branch is unicameral and consists of 21 senators, each popularly elected. Guam also has a judiciary branch. Two local courts (a Superior Court and a Supreme Court) and one U.S. District Court serve the island. Local judges are appointed by the governor and are confirmed by voters every 6 years (DOI, 1996b).

Economy

Compared with the other U.S.-Associated Pacific Basin jurisdictions, Guam enjoys a healthy economy. The gross island product (GIP) for 1994 was $3.011 billion, a per capita GIP of $20,640, the highest of the U.S.-Associated Pacific Basin jurisdictions (Bank of Hawaii, 1995b). Primary sources of revenue are U.S. federal expenditures and tourism. In total, U.S. federal expenditures (military, grant assistance, and other payments) account for about one-third of Guam's revenue.

Employment

Guam's largest single employer is the government of Guam (GovGuam). On average GovGuam pays almost twice as much as the private sector (Guam Bureau of Planning, 1996). GovGuam employed 20.2 percent of the workforce (13,430 individuals); the U.S. federal government employed 10.4 percent of the workforce (6,390 individuals). The private sector—primarily the services and tourism industry—employed most of the remaining 46,100, or 69.4 percent (Bank of Hawaii, 1995b).

Federal Expenditures

Federal Grant Assistance

Total federal grant assistance to Guam in Fiscal Year 1996 was $134 million, most of which came from the U.S. Department of the Interior (44.6 million) and the Department of Health and Human Services ($19.8 million) (Bureau of the Census, 1997). GovGuam also directly receives federal income taxes paid by residents; this money goes directly into the local budget.

Military Spending

Total annual spending for the military in Guam in 1994 was $750 million. The U.S. military on Guam operates Navy and Air Force bases and employs about 7,000 civilians (Bank of Hawaii, 1995b). Average salaries for these civilians are double those in the civilian economy, making the jobs much more attractive. Some downsizing of the military has begun, and the military projects the loss of 1,100 to 1,200 jobs between 1996 and 2001. Even so, military spending is not expected to drop dramatically between 1996 and 2001 (Bank of Hawaii, 1995b).

Tourism

Guam is a desirable destination for travelers from Japan and East Asia. In the mid-1980s it experienced a tremendous growth in visitors to the island, from 368,620 visitors in 1984 to 780,404 visitors in 1990. Well-known for its good infrastructure and services, Guam has responded well to increasing numbers of visitors by upgrading its facilities, including a new airport terminal and new hotels and plans to upgrade some of its basic infrastructure, like roads, power, and water services. Tourism will likely continue to grow and emerge as the primary source of income (Bank of Hawaii, 1995b).

Population

The first settlers of Guam and the Mariana Islands, Chamorros, are believed to have originated in Southeast Asia. Today's Chamorros are a mix of Chamorro and Filipino. While the numbers of Chamorros in Guam have increased, they represent a lesser proportion of the total population than in the past. Prior to WWII, Chamorros represented about 90 percent of the civilian population, but today they represent less than half (43 percent) of the civilian population (GHPDA, 1996; PIHOA, 1997). The remainder are Filipino (23 percent), Caucasian (14 percent), other Asian (14 percent), and other Micronesian (6 percent) (GHPDA, 1996; PIHOA, 1997). About 145,000 (80 percent) are local

APPENDIX D

residents and about 20,000 (20 percent) are military personnel, dependents, and retirees (of which about 7,000 are active-duty personnel) (Bank of Hawaii, 1995b).

Guam's annual population growth rate of 2.6 percent is the lowest growth rate in the region (PIHOA, 1997). Nonetheless, if this growth rate continues, Guam's population will almost double by 2050 (U.S. Bureau of the Census, International Database, 1997b). In the last decade, Guam's tourism boom has attracted immigrants from Southeast Asia (primarily the Philippines) and the Pacific in search of job opportunities (Bank of Hawaii, 1995b; East-West Center, 1996). Some of the migration has been from the freely associated states, particularly after the Compacts of Free Association granted citizens from the freely associated states unlimited access to the United States and its territories. The number of immigrants from the freely associated states has increased steadily since the implementation of the Compacts, and in 1994 there were about 6,630 FSM-born, and a few hundred RMI- and Palau-born, residents (Levin, 1996).[9]

Infrastructure

In general, Guam's has the most stable infrastructure of the U.S.-Associated Pacific Basin jurisdictions. GovGuam owns and operates all public utilities: telephone, power, water, and sewer systems (Bank of Hawaii, 1995b). Some concerns remain, however. One of the biggest safety concerns is the frequent failures of the power system, operated by the Guam Power Authority (GPA). During site visits, several people reported that the agency has been plagued with mismanagement and that in the past brownouts occurred on a frequent basis. Some of the problems have been explained by the rapid increase in demand, especially during the spurt of tourism in the late 1980s, and increases in residential power demands. Brown tree snakes, which frequently climb electrical lines, have also been held responsible for outages. GPA has hired an off-island management firm to overhaul the electrical system, and reportedly outages are less frequent.

Although the water is considered potable, the Power and Utilities Authority of Guam (PUAG) has also had difficulty with the water system; about 30 to 40 percent of the water goes unaccounted for each day. Persistent leakage and unmetered use are primarily responsible (DOI, 1996b). In addition, it was reported during the site visits that many of Guam's public water wells have greatly reduced fluoride levels, raising a concern for dental health. It was further

---

[9]Data were compiled from several sources by the Department of the Interior, Office of Insular Affairs for its report to Congress, *The Impact of the Compacts of Free Association on the United States Territories and Commonwealths and on the State of Hawaii*, September 1996. Please note data are estimates only; they were compiled from different surveys and are not current. Therefore, the numbers may reflect a phenomenon that may or may not still be true today.

reported that water on Guam has high levels of lead and other minerals, such as manganese, which is another public health concern.

HEALTH CARE DELIVERY SYSTEM ORGANIZATION

Administration

Government health care services on Guam are organized into four independent agencies of GovGuam: Guam Memorial Hospital Authority (GMHA), Department of Public Health and Social Services (DPHSS), Department of Mental Health and Substance Abuse (DMHSA), and the Department of Vocational and Rehabilitative Services. In addition, Guam has several private health clinics.

Health Care Facilities

Hospitals

Guam has two hospitals. The Guam Memorial Hospital (GMH) serves civilians, and the Naval Hospital serves active-duty military personnel, their dependents, and retirees. Guam's two hospitals, two federally licensed health maintenance organizations (GMHP and FHP, see below), and DPH&SS all must adhere to federal quality guidelines to receive payment (GHPDA, 1996).

Guam Memorial Hospital Authority GMHA operates GMH. With final expansion and construction completed in 1991, the hospital has 192 beds, comprising 159 acute-care beds and 33 long-term-care beds in the skilled nursing facility (GHPDA, 1996). All hospital services are provided at one campus. GMH is run by a board of directors and an administrator, all of whom are appointed by the governor and confirmed by the legislature. The following are some of the issues of concern to GMHA:

Financial In the past GMH has been plagued with an inadequate billing system and a poor ability to collect fees. It was reported during the committee's site visits that although the hospital implemented a new computerized billing system, many staff have not been thoroughly trained, the error rate is high, and there is only one computer system support staff. In addition, the billing format does not conform to Medicaid specifications, which results in delayed payments. Other reasons cited for low rates of fee collection are the number of self-payers who do not pay their bills. In 1994 the hospital collected 72.2 percent of the money that it was owed. The rate is only slightly increased from that in 1990 (GHPDA, 1996).

Privatization The governor is pursuing a plan to privatize GMH, transferring partial ownership to community members willing to invest capital in the hospital. The governing board would consist of government representatives and investors. The increase in revenue would be used to upgrade the facility.

Accreditation In 1990 the Health Care Financing Administration (HCFA) cited GMH for numerous deficiencies (113 altogether), including the use of unlicensed personnel, lack of accountability for quality of care and the safety of patients and staff, building maintenance, shortages of staff, and the lack of complete records. GMH has since corrected many of the problems, and in 1995 it received only 22 citations (HCFA, 1997). Nevertheless, GMH is still pursuing, although it has not re-attained the Joint Commission on the Accreditation of Healthcare Organizations accreditation that it lost in 1983 (GHPDA, 1996).

U.S. Naval Medical Regional Center (USNMRC) The military health system on Guam consists of the Naval Hospital and several smaller dispensaries for general acute care. USNMRC provides outpatient services and runs one small dental clinic. The center is self-contained and is staffed and equipped to serve primarily active duty military, their dependents, and retirees. Limited coordination between the Naval Hospital and GMH exists, however. The Naval Hospital accepts emergency cases to stabilize patients and serves as a backup to GMH during natural disasters and other island emergencies, such as the recent crash of a South Korean passenger jet. Provisions in the Compacts direct the Defense Secretary to make all U.S. Department of Defense medical facilities available to citizens of the freely associated states, who are "properly referred to such facilities by government authorities responsible for the provision of medical services" (P.L. 99-239), although this does not appear to occur very often.

Public Clinics

Department of Public Health and Social Services DPHSS operates four regional health centers. One is funded through a federal Community Health Center Grant; the others are funded by DPHSS local and other grant monies (GHPDA, 1996).

The following programs and services are available at the regional health centers:

- maternal and child health,
- family planning,
- nutrition and health program (Women, Infants, and Children, or WIC),
- limited medical support services,
- services for children with special health care needs,
- communicable disease control, and
- dental services.

Department of Mental Health and Substance Abuse DMHSA has a 48-bed inpatient facility with both a medical services division and a clinical services division (16 beds for residential treatment for drug and alcohol abuse, 16 beds for adults, and 16 beds for children). There is some concern that with increases in cases of substance abuse, especially ones that involve drug combinations (e.g., alcohol and methamphetamines), and violence, the facility may not be big enough to handle the demand for services and will need staff better trained to cope with the situation (GHPDA, 1996).

Department of Education Each of Guam's 24 elementary, 6 middle, and 5 public high schools has a nurse's station staffed by one full-time registered nurse (some also have a licensed practical nurse) to provide health services to students and to refer children to maternal and child health services when needed.

Private Clinics

Guam has a total of 22 private health clinics (PIHOA, 1997), including multispecialty Seventh Day Adventist and FHP clinics (nonprofit), The Doctors' Clinic (multispecialty and for profit), and various other for-profit private practice clinics and auxiliary services. There are currently three home care agencies on-island: Guam Nursing Services (HCFA-approved); TropiCare (serving primarily the FHP/CHOICE PLUS health plan enrollees); and Micronesian Home Health Care (serving mainly GMHP health plan clients). There is also one non-profit, Catholic long-term care facility, Saint Dominic's Senior Care Home.

Health-Related Community Organizations

Guam has a rich variety of health-related community organizations including the American Red Cross, American Cancer Society, Lytico-Bodig Association, Hemophilia Association, and Guam Diabetes Association. Hospice Guam is another non-profit community organization that supports terminally ill persons and their families. Further, Guam is in the process of organizing a chapter of the Arthritis Foundation.

Guam, like its neighbors, sends patients off-island for diagnoses and for treatments that are not available locally. Unlike the other islands, however, the decision to refer a patient for off-island tertiary care is left mostly to the doctor in charge of the patient's care and does not rest with a political entity. Some of the private health plans have agreements with hospitals in Hawaii and southern California for referrals.

Lack of equipment and specialty services for diagnosis and treatment are the most likely reasons that people are referred off-island. Typical services for which patients are referred include cardiac procedures (e.g., bypass surgeries), neonatal treatment, and radiation treatment for cancer (although there is a new Cancer Institute of Guam, complete with a radiation center, that should help to lessen the need to seek chemotherapy treatments off-island). Because malpractice insurance for providers does not exist on Guam, some speculate that doctors are more likely to send people off-island for diagnoses that they would otherwise make if they were insured against lawsuits (GHPDA, 1996).

Because a number of off-island tertiary care referrals are self-initiated or are paid for through private health insurance, costs attributed to off-island referrals are difficult to estimate. GovGuam, through the Medically Indigent Program (MIP) and Medicaid spent about $4.9 million on off-island tertiary care referrals for 163 people in 1996 (T. Fejeran, personal communication, October 20, 1997). Unfortunately, no uniform system of data collection or mandated reporting exists among the private providers, so accurate data in the numbers of people referred and the costs incurred are not available (GHPDA, 1996).

HEALTH CARE RESOURCES

Financial

Similar to their neighbors, Guamanians believe in and practice the right of health care for all, regardless of the ability to pay. Unlike the other Pacific jurisdictions, however, the private insurance market is well established, with growing numbers of group health entities and HMOs, similar to the situation in small markets in the United States. Although these private insurers cover much of Guam's health care costs, including off-island referrals, the health care costs for many who cannot afford private insurance (about 10 to 15 percent of the population) are the government's obligations (see below).

Compact Impact

Guam has become increasingly concerned over the impact that the Compacts have had on migration citizens of the freely associated states to Guam (called "Compact impact"). Estimates indicate that between 1988 and 1992 (2 years after the Compacts became effective), the numbers of residents born in the FSM and residing in Guam increased almost three-fold: from 1,700 in 1988 to 4,954 in 1992 (Levin, 1996).[10] The Guam Department of Public Health and

---

[10]Data were compiled from several sources by the U.S. Department of the Interior, Office of Insular Affairs for its report to Congress, *The Impact of the Compacts of Free Association on the United States Territories and Commonwealths and on the State of Hawaii* (DOI, 1996). Note that data are estimates only; they were compiled from different surveys and are not current. Therefore, the numbers may reflect a phenomenon that may or may not still be true today.

Social Services estimates it spent $3.7 million between 1989 and 1992 for health services provided to citizens of the freely associated states (Government of Guam, 1993). In 1995, Guam received $2.5 million for impact costs, plus technical assistance and other grants aimed at measuring those costs (DOI, 1996b). Many citizens of the freely associated states, however, find jobs on Guam, have health insurance, or can pay for the costs of their own care, and pay taxes which offset local government expenditures.[11]

Public Health Insurance

Public health insurance on Guam includes federal Medicare and Medicaid programs, but it also includes an entirely locally funded safety net program that covers the under- and uninsured (MIP). The Comprehensive Health and Medical Plan for the U.S. Armed Forces (CHAMPUS) also provides insurance for active-duty military personnel, their dependents, and retirees.

Medicare enrolls about 2,000 seniors and about 8,000 people receive Medicaid (GHPDA, 1996). Guam receives federal funding for its Medicaid program at a 50–50 matching rate, but with a cap on federal assistance of $4.06 million (Guam Legislature, 1997). With growing numbers of enrollees and increasing costs, GovGuam has had to pay as much as 85 percent of its Medicaid costs (GHPDA, 1996). The Medicaid program is administered through DPHSS.

For the approximately 16,000 to 21,000 (or about 10 to 15 percent) of Guam's population who are uninsured and who are not eligible for Medicare or Medicaid, GovGuam provides coverage through MIP (Guam Task Force, 1995). Administered by DPHSS and funded entirely through local money, MIP covers the costs for treatment and services for people whose incomes do not meet federal criteria, but who cannot afford health insurance premiums and people who have used up all of their available private health insurance benefits. MIP also acts as a supplement for the underinsured and Medicare recipients. Current concerns are that MIP cannot continue to function unless it is given more resources or reduces its expenditures. The average cost per recipient increased from $1,492 in 1991 to $3,663 in 1995, a 26 percent average increase per year (Guam Legislature, 1997).

Recently, legislation was introduced in the Guam Legislature to contract HMOs to provide services both for MIP participants and Medicaid enrollees. This move would presumably not only cut expenses, but would also allow Guam to apply for a Medicaid Section 1115 waiver, potentially allowing it to garner more federal money to supplement the program (Guam Legislature, 1997).

---

[11] For a more complete discussion of "Compact impact" and references to relevant literature, the reader is referred to Levin, 1996, "Micronesian Migrants to Guam and the Commonwealth of the Northern Mariana Islands: A Study of the Impact of the Compact of Free Association."

Private Health Insurance

Most of the people in Guam receive their health insurance through their employers. Eighty-two percent of Guam's population is covered by some form of health insurance: for about three-quarters of the people employers provide coverage through 1 of the 22 private health plans; the remainder are covered through one of the public programs (PIHOA, 1997). Another small percentage of the population can and does pay out of pocket for services. GMHP, which was once owned by the government, is a private HMO and, together with FHP/CHOICE PLUS HMO, covers about one-half of the privately insured population (GHPDA, 1996). Other plans include Staywell, HML, and Multicover. In addition, some of Guam's private health insurers have expanded services to the Commonwealth of the Northern Mariana Islands (CNMI), broadening the insurance market.

Workforce[12]

Guam, like most islands or rural locations, suffers from isolation and the difficulties with the recruitment and training of a local workforce this incurs. To deal with such a shortage of trained personnel, Guam has flown in physician specialists for temporary work and has employed workers from other countries (many of whom are from the Philippines and work at GMH). Guam has also used international recruiting methods, such as advertising through the Internet and U.S. professional journals.

Physicians

Guam has a total of 311 civilian M.D.s, or 1 physician for every 500 people (PIHOA, 1997). Many of these physicians are contract workers, primarily from the Philippines, as is the case for CNMI.

Nurses

Currently, Guam has 648 registered nurses and 164 licensed practical nurses on-island, according to licensing records (PIHOA, 1997). In 1995, there were no nurse's assistants, 166 at GMH, and the remainder with the DPHSS (GHPDA, 1996). Although Guam may have a reasonable supply of registered nurses and licensed practical nurses, serious problems with recruitment and retention have led to shortages of nurses at GMH and public health clinics. At

---

[12] Workforce data are from the 1997 PIHOA Data Matrix (PIHOA, 1997).

GMH, long work hours, shortages of other staff, and little continuing education and support have been cited as reasons that nurses are leaving.

Because GMH is unable to recruit staff on-island, it has hired many non-U.S.-trained medical staff (many are Filipino) and has traditionally allowed them to work as mid-level practitioners until they passed the U.S. board examination. In 1990 HCFA cited GMH for using unlicensed personnel (HCFA, 1997). The committee learned at the site visits that after the citation many staff were let go. The committee was also told that the hospital fired 37 Filipino M.D.s and that as many as 90 percent of the Filipino-trained registered nurses were reclassified as nurse's assistants.

Other Health Care Personnel

A variety of allied health care personnel provide support and auxiliary health care services on Guam. Seventy dentists, 21 dental hygienists, and 221 dental aides serve the island, and most of these work in private practices. DPHSS has 26 dentists (one is a National Health Service Corps dentist) who work in the three regional health centers. Other personnel include 57 pharmacists and 13 laboratory technicians. Guam is lacking trained technicians such as radiologists, laboratory technicians, and assistants. However, Guam Community College is in the process of resurrecting its allied health program to develop a supply of technicians and other auxiliary personnel.

Traditional Health Practitioners

Traditional forms of medicine include the use of massage, herbs, coconut oil, and prayer and are administered by *suruhanus*, traditional Chamorro healers. Although their practices are well known in communities in Guam, they are not integrated into the health care system, and sometimes they are referred to as "witch doctors."[13] However, many people turn to traditional medicines when Western medicine offers no hope of cure or relief from pain. People consult *suruhanus* for conditions such as infertility and stomach ailments. *Suruhanas* work with the medical establishment insofar as they ask patients to see a doctor first and to inform them of any medications they are taking. In addition, during the site visits the committee was told that patients at GMH are permitted to call in *suruhanus*.

---

[13] Witch doctors, or *kakahna*, were believed to have supernatural powers to create and cure illnesses, but they are different from *suruhanus*, whose practice is limited to using various herbs and massage to treat illnesses.

APPENDIX D   139

Workforce Quality Assurance

There are two licensing boards on Guam for physicians and nurses. The Guam Medical Association and Guam Nurses Association are both very active. HCFA accredits the hospital, skilled nursing facility, and home health agency.

Technology, Supplies, and Equipment

On Guam, there is some debate about whether more money should be used to purchase or upgrade needed equipment so that some diagnosis and treatment services can be provided on-island. For example, Guam had no magnetic resonance imager (MRI) on-island at the time of the site visit (although it reportedly does have one now), and no radiation treatment is provided. In addition, with the island's high numbers of dialysis patients and only one dialysis unit at GMH, and only one smaller private dialysis treatment center, there is a demand for increased numbers of dialysis machines. In July 1995 GMH's dialysis unit was providing treatments to 102 patients (GHPDA, 1996). Some have argued that efforts should focus on providing funding for kidney transplants, which would eliminate the need for a lifetime of dialysis and would thus save much more money in the long run.

On the computer technology front, Internet service providers have begun to enter Guam's market. All public libraries, schools, and many homes now have access to the Internet. Since Guam was connected to the U.S. telephone network in July 1997, the cost of long-distance service has been dramatically reduced, making Internet access much more feasible. During the site visit, however, few trained computer services personnel were available to maintain the equipment. The committee visited one public library with three computers, all with Internet access, but only one machine was working, and therefore, access was limited to 1 hour per person. GMH has one computer located in a small library for use by hospital staff. Although hospital staff have expressed a desire to have Internet services, at the time of the committee's visit such services were not available.

FUTURE HEALTH CARE ISSUES

With a private health care system already well established, Guam's attention and hopes for its future rest with public–private partnerships. The current governor has initiated the Vision 2000 Campaign for Guam. The aim of this campaign is to establish priorities in all areas of the island's government, social services, infrastructure, and environment. In health care, the governor has particularly emphasized promoting Guam as a regional health center for the western Pacific (Government of Guam, 1997). Toward that end, some of the following are top health service priorities:

- privatize GMH;
- focus efforts on disease prevention, especially diabetes and cardiovascular diseases;
- increase the numbers of physicians, and increase the amount of available specialty care services, equipment, and technology so that more can be provided on-island and so that Guam can be more widely used for regional referrals; and
- make health care accessible and affordable by controlling costs (e.g., contracting out Medicaid and MIP to private managed care) and enhancing public health programs and services that can reduce the need for acute care.

Republic of the Marshall Islands

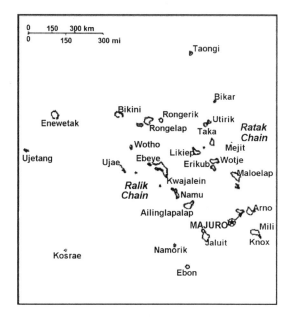

Total Population	59,246
Number of Inhabited Islands and Atolls	23
Access to Major Health Facility (percentage of population requiring more than 1 hour of travel)	53%
Total Health Budget	$7,600,000
Per Capita Health Budget	$128

SOURCE: PIHOA (1997). NOTES: Total population is the official estimate from the RMI 1996 census. The total health budget excludes money for people exposed to radiation.

OVERVIEW

The people of the Republic of the Marshall Islands (RMI) live among 23 islands and coral atolls, tips of ocean volcanoes that have long since receded beneath the water. Although it is now an independent and sovereign republic, RMI remains closely involved with the United States. This involvement began after World War II when the United States took control over the islands from the Japanese. Beginning in 1946 and continuing until 1958, the U.S. Navy conducted nuclear tests in the region. As a result, hundreds of Marshallese people were relocated from their ancestral homes, and 253 Marshallese are known to have been directly exposed to radioactive fallout. Concerns about the effects of the nuclear testing are ever present in RMI, even today, nearly 40 years after the last nuclear test.

Government

RMI has a unicameral government, a combination of U.S. democratic and British parliamentary systems. The Nitijela, or parliament, is the legislative body and comprises many committees. The Council of Iroij (chiefs) may provide opinions to the Nitijela on any matters concerning the nation, and they may also ask for reconsideration of any proposed bill. Any Marshallese citizen older than 18 years of age may run for a seat on the Nitijela, whereas seats on the Council of Iroij are passed down through families. There are two courts: a Supreme Court, with judges appointed by the RMI President's cabinet, and a High Court. Marshallese are a matrilineal society, with very organized land holding structures, and so there is also a Traditional Rights Court that deals with legal questions surrounding land and traditional practice (DOI, 1996b).

Economy

The Marshallese economy is almost entirely dependent on foreign aid, the vast majority of which is from the United States, either in the form of money from the Compact of Free Association or in the form of rent paid for the use of a missile testing facility on the atoll of Kwajalein known as USAKA (United States Army-Kwajalein Atoll). A very small amount of tourism and trading in copra (dried coconut meat) provide some local revenue.

As with other jurisdictions, the local government employs the greatest number of people, about 34 percent of the labor force (Department of State, 1994). USAKA is another important employer, providing jobs to roughly 1,500 Ebeye residents (Bank of Hawaii, 1996). Aside from local government and U.S. military civilian jobs, there is little private industry. Overall, there are a very limited number of jobs to absorb the growing numbers of people entering the labor force each year.

Population

RMI had one of the world's highest population growth rates: 4.1 percent annually between 1980 and 1988 (Bank of Hawaii, 1996). Although it decreased somewhat in 1997 to 3.9 percent, the population is still expected to double in the next 20 years (U.S. Bureau of the Census, 1997a). The total fertility rate for a Marshallese woman is 7.2 children (World Bank, 1994). Population growth is one of the primary social concerns in the Marshall Islands and an issue that has been given top priority in government planning (Republic of the Marshall Islands, 1990).

In addition to high growth rates, the RMI population is also a young one and is the youngest of the six jurisdictions of the U.S.-Associated Pacific Basin. Half of the RMI population is under the age of 15 (PIHOA, 1997). About 20 percent of the births in 1994 were to teens under the age of 15 (HRSA, 1996).

Population density is concentrated in two atolls: Majuro and Kwajalein. About half of the total population lives on Majuro Atoll; most of these people live in the 0.51 square mile of the Djarrit-Uliga-Dalap (D.U.D.) area, which has a density of 28,724 persons per square mile. Even more densely populated is Kwajalein's Ebeye Island: with a population of 8,324 persons in 1988 and a land area of only 0.14 square miles, it had a density of 59,457 persons per square mile—one of the highest densities in the world (Republic of the Marshall Islands, 1990). With such extreme population densities in its urban centers, the ensuing environmental and health problems are of great and urgent concern.

Infrastructure

One of the most significant public health risks to RMI is the lack of potable water and a very poor sewage system. Rapid population growth, notably in Majuro and Ebeye, poses serious threats to public health. In 1995, less than a quarter of the households in RMI were connected to a public water system, and according to the World Bank, between 1983 and 1985[14] only about 31 percent of the population had access to safe drinking water (World Bank, 1994). At the time of the committee's site visit, water had been cut off in Ebeye for almost a week.

Communications within and among the atolls pose challenges. Most people outside Majuro and Ebeye do not have telephones or electricity. Although more people are obtaining Internet access, this is primarily limited to the College of the Marshall Islands (CMI) through PEACESAT, and a few others who have obtained access through private providers. Only a little more than half (54 percent) of the households had electricity in 1995 (DOI, 1996b), and power reliability is a problem. For example, during the site visit, electricity on Ebeye was available on a rotating basis with some areas receiving service during the evening and others receiving service during the day.

HEALTH CARE DELIVERY SYSTEM ORGANIZATION

Administration

The health care system is administered and subsidized by the Marshallese government through the Ministry of Health and Environment (MOH), a cabinet-

---

[14] Although there may have been some improvement since this time, these estimates remain fairly close to what exists today.

level agency. Within MOH, the secretary of health oversees four major departments: Primary Health Care, Kwajalein Atoll Health Care, Majuro Hospital, and Administration and Finance. In addition, MOH administers a special program known as the 177 Health Care Program, which provides health care to individuals adversely affected by nuclear testing. The U.S. Department of Energy (DOE) administers yet another program, the Marshall Islands Medical Program, specifically for people who were directly exposed to nuclear fallout. Both of these programs are explained in more detail below.

Marshall Islands Medical Care Program

As mandated through the Compacts of Free Association, the Marshall Islands Medical Care Program provides medical care and treatment for potentially radiogenic diseases for the remaining exposed victims[15] of the 1954 Bravo test on Rongelap and Utrik plus screening and acute care for a comparison group of 109 people. This special medical care and surveillance have been provided by the Brookhaven National Laboratory via a contract with DOE. The program receives annual funding from the U.S. Congress at about $2 million (Bell, 1997).

Under the current structure of the Marshall Islands Medical Care Program, teams of doctors are sent to RMI for 1-month missions two times per year. The mission teams provide full medical examinations, including thyroid and endocrine examinations, gynecological examinations, various urine and blood examinations, and diagnostic tests that can include mammograms, thyroid ultrasound, and X-rays. Patients with conditions potentially radiogenic in origin are referred to off-island facilities (primarily Hawaii) when diagnoses and treatments for those conditions are not available in RMI. Patients requiring care for other conditions and between missions (and patients in the comparison group who need medical attention) are referred to the 177 Health Care Program. The structure of the Marshall Islands Medical Care Program is in the process of being changed to place a greater emphasis on providing more holistic and community-based care to the individuals served by this program (Bell, 1997).

Section 177 Health Care Program

The Compact of Free Association also provided money for a "Four Atoll Health Care Program" to provide health care services to people of the four atolls affected by U.S. nuclear testing (Bikini, Enewetok, Rongelap, and Utrik), as well as to their descendants. Named for the section of the Compact that speaks to nuclear testing effects, the 177 Health Care Program provides care for

---

[15] A total of 253 people were directly exposed to fallout from the Bravo tests in 1954. However, only 131 are still alive.

approximately 11,470 Marshallese people. The U.S. Department of the Interior annually provides about $2 million to the Marshallese government for the program, and RMI government contracts administration of the program to a private health care organization (Bell, 1997).[16] To be eligible for the program a person must be Marshallese and must have been residing on one of the four atolls during 1946–1958 or be a direct descendant of a resident.

Health-Care-Related Community Organizations

The Youth to Youth in Health Program is a nongovernmental organization that promotes youth and community involvement with primary health care. The program trains young people to be peer educators and to serve as role models. It sponsors health outreach programs and produces radio and television programs promoting health and cultural awareness on such topics as nutrition, family planning, substance abuse, and mental illness. In 1995, a health clinic for adolescents opened in Majuro. Another aspect of the project is to promote and supervise income-generating projects for outer islanders. The program operates on 20 islands and receives some support from MOH (Youth to Youth in Health, 1996).

Off-Island Care

RMI spent approximately 33 percent of its total health care budget serving 148 patients in 1996 (PIHOA, 1997). An off-island referral committee is chosen by the secretary of health. However, during the committee's site visit it was reported that about 60 people are on the waiting list to be sent off-island for medical care, with the government able to afford sending only 1 or 2 each month. Consequently, there are also reports that decisions of who is sent off-island are based largely on political favors and social status, with no internal consistency in the criteria or decisionmaking used to refer people off-island. Added to the problem is the higher level of attention and financial support given to radiation-related health problems; those patients are often referred to hospitals in Hawaii (a preferred location) or the Philippines.

Health Care Facilities

RMI has two major hospitals, located in the major urban centers of Majuro and Ebeye. Built in 1986, the Majuro Hospital has 103 inpatient beds, an emergency room, and a dental clinic. The structure itself is largely constructed

---

[16]Until the fall of 1997, Mercy International Health Services administered the Section 177 Health Care Program.

with specially coated cardboard paneling. MOH estimates the hospital can only last another 5 years and would like to replace it as soon as possible. The Ebeye Hospital has 25 inpatient beds and is in serious disrepair. A new hospital facility has been built with funds from the Department of the Interior but it remains unoccupied because there are no funds to purchase and install the equipment necessary to make it operational. The U.S. Army hospital on the USAKA base, which serves military personnel, their dependents and, very occasionally, some Marshallese individuals who work at the base, is in close proximity but is not available to the general Marshallese population.

A Health Resources and Services Administration (HRSA)-funded Section 330 community health center operates in Ebeye. The Youth-to-Youth in Health Program also operates a health clinic for adolescents in Majuro as well. Among the outer islands and atolls are 58 dispensaries. Currently, each is run by a health assistant and is linked to the Majuro Hospital through shortwave radio. Many of the health assistants are nearing retirement. However, the government recently implemented a health careers opportunity program (see below) to train a new set of health assistants at the Majuro Hospital.

HEALTH CARE RESOURCES

Financial

The total health care budget for RMI for Fiscal Year 1996 was $7.6 million (PIHOA, 1997). Funding for operations comes from Compact money, the general fund, U.S. funds for primary health care and public health, and other grants. The universal health care system (the Marshall Islands Health Plan) provides for and insures every Marshallese resident. User fees are charged for health services, but the fee amount itself is low (e.g., $2 per outpatient visit) and the collection effort is minimal. Radiation-exposed victims, their descendants, and current residents of the four atolls exposed to radiation are insured separately under the Marshall Islands Program and the Section 177 Health Care Program, funded by Compact money.

Workforce[17]

Physicians

The RMI relies heavily on expatriate physicians; 13 of the 19 M.D.s working in RMI are expatriates (9 are from the Philippines and 1 each from the United States, Australia, Burma, and Sri Lanka). There is also one expatriate M.B.B.S. One of the hospital doctors has reportedly opened a private practice.

---

[17]Workforce data are from the 1997 PIHOA Data Matrix (PIHOA, 1997).

Currently, six graduates of the Pacific Basin Medical Officers Training Program (PBMOTP) are working in RMI (five are indigenous, one is from the Federated States of Micronesia; one more is on maternity leave, but she plans to practice in RMI). These medical officers provide staffing for the two hospitals' inpatient and outpatient units and emergency room. With only one Marshallese student known to be enrolled in medical school, RMI will remain dependent on expatriates for the foreseeable future.

Dentists

All four dentists in RMI are expatriates (two are from Burma, one each is from the United States and the Philippines), three are located on Majuro, and one is located on Ebeye. They are assisted by four dental nurses and eight aides. At least two Marshallese students are enrolled in the school of dentistry in Fiji.

Mid-Level Practitioners

Nine medexes and 56 health assistants staff the dispensaries. However, the average age of the health assistants is 56, and many plan to retire in the near future. MOH has recently begun to offer a health care opportunities program to train new health assistants. The program, funded by a HRSA grant, enrolls high school graduates in a special 18-month training program that includes 4 months of formal classes at MOH and 5 months of work in the Majuro Hospital before they are sent out to staff the dispensaries.

Nurses

The College of the Marshall Islands (CMI) is an accredited 2-year college and offers an associate degree in nursing. Nurses from CMI staff the hospitals not only in RMI, but also throughout the region. However, with a relatively low current enrollment, concerns are mounting that the hospitals will soon experience a shortage of nursing personnel. CMI is actively seeking to expand its nursing program to offer a 4-year bachelor's degree program. Accordingly, although the island has only 1 registered nurse, 79 graduate nurses (96 percent of whom are indigenous) and 44 nursing aides are in practice.

Traditional Health Care Practitioners

In the Marshall Islands there has been no tangible carryover of pre-Western traditional healers, as is the case with the *suruhanos* in Guam, for example. However, particular cultural beliefs about family, gender roles, privacy, and religion do influence health behavior, and at times, these beliefs place people at odds with the Western medicine that so dominates the system.

Technology, Supplies, and Equipment

A chronic lack of critical supplies and equipment exists. In recent years, the Army Hospital on Kwajalein has taken some referrals and has used telecommunications technology to send images to doctors at Tripler Army Medical Center in Hawaii for diagnoses, although it has primarily been for dermatological patients (Bice et al., 1996).

FUTURE HEALTH CARE ISSUES

With rapid population growth, child malnutrition, and increases in the occurrence of diseases resulting from lifestyle factors, such as diabetes, RMI is recognizing the need for restructuring its health care system to focus on primary care. In 1987, the RMI government launched a national campaign to focus on primary health care and to involve the community in planning services. To this end the government is in the process of building 21 community health stations that will be operated and maintained by community health councils (which will include community leaders, church leaders, and residents), staffed by health assistants, and regularly visited and assisted by primary health care teams consisting of doctors, nurses, dentists, and social services staff.

The primary health care team has already begun its work on Ebeye with 10 community health councils. At the time of the committee's site visit, the team had completed the process of collecting data such as household characteristics, and level of access to public water, power, and sewage systems to establish community profiles. This primary health care team appears to be highly motivated about its work, communicated effectively, and was knowledgeable about residents' needs. It is hoped that this primary health care approach, with an emphasis on education, will empower people to better care for their own health and the health of the people in their community.

Palau (Belau)

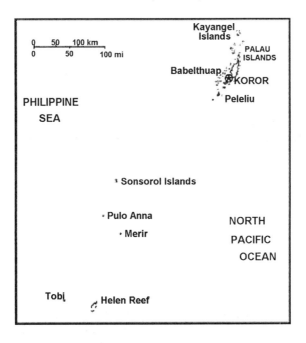

Total Population	17,225
Number of Inhabited Islands and Atolls	8
Access to Major Health Facility (percentage of population requiring more than 1 hour of travel)	70%
Total Health Budget	$10,912,500
Per Capita Health Budget	$633

SOURCE: PIHOA (1997). NOTE: Total population is the official estimate from the 1990 U.S. census; health care budget is from Fiscal Year 1997.

OVERVIEW

Palau became an independent country in 1994 after years of negotiations with the United States and several national plebiscites. Like Guam and the Commonwealth of the Northern Mariana Islands (CNMI), Palau is currently experiencing somewhat of an economic boom, with marked increases in tourism, new construction, and development and, relatedly, an increase in foreign workers—primarily from the Philippines. At the same time new emphasis has been placed on the promotion of primary health care and preventive services. A new hospital has been constructed, and several superdispensaries have been established. Much of the primary care is being provided by newly graduated medical officers from the Pacific Basin Medical Officers Training Program (PBMOTP).

Government

Palau has several layers of government. At the national level, there is a president, vice president, and a national congress with 14 senators and 16 delegates. The national government also receives advice from a council of chiefs, formed of one traditional chief from each state. At the state level, each of the 16 states has a governor and a state legislature.

Population

Although it is estimated that the Palauan Islands may have been home to as many as 45,000 people in 1783 when westerners first landed and more than 30,000 people during the height of Japanese administration in the 1930s, Palau's current population is only about 17,225 (PIHOA, 1997). Only 8 of Palau's more than 200 islands are inhabited, and 70 percent of the people live on the island of Koror. More than 80 percent of the total population is native Palauan; the ethnicity of the rest of the population is mostly Filipino. Fewer than 100 people live in Palau's Southwest Islands; ethnically, they are considered Southwestern Islanders rather than Palauan.

Economy

Palau's economy is based primarily on tourism and government expenditures. In 1995 approximately 65 percent of the labor force worked in the private sector, whereas the remaining 35 percent was employed by the government (DOI, 1996b). Palau's economic future appears bright, with confident forecasts of increased tourism and resulting development in the coming years (Bank of Hawaii, 1994). In addition, unlike the Compacts of Free Association with the Federated States of Micronesia and the Republic of the Marshall Islands, Palau's Compact with the United States provides for a trust fund for long-term capital investments.

Infrastructure

Although nearly all housing units in Palau (92 percent in 1995) have access to a public water source, the water remains unsafe. According to the U.S. Department of the Interior (1996b), the water treatment plant that serves the majority of the population on Koror does not meet U.S. Public Health Service standards for public water systems. In the outlying areas and islands, people must rely primarily on rain water catchment systems, surface water sources, or shallow wells to meet their needs (DOI, 1996b).

Inadequate sewage and waste disposal systems present another hazard. Only about 41 percent of households in 1995 (up from 30 percent in 1990) were connected to a public sewer system. The only waste treatment plant has reached capacity, and trash collection has been infrequent, causing a buildup of garbage in and around residences. Coastal waters and harbors are beginning to show signs of contamination (DOI, 1996b).

HEALTH CARE DELIVERY SYSTEM ORGANIZATION

Administration

The minister of health, a cabinet-level appointee, administers the overall health care system. Under the minister are a director of the Bureau of Public Health and Primary Health who manages all outpatient activities, dispensaries, and superdispensary services as well as all other federally funded health services and programs and a director of the Bureau of Clinical Services who manages medical inpatient activities with the Belau National Hospital. Although most health care is provided through the government, a small and growing private medical practice has been established.

Off-Island Care

A team of senior physicians must make a majority decision about any recommendation for off-island tertiary care referral. Most of the patient referrals (75 percent) are to the Philippines because the facilities and services there are closer and less costly than those in Hawaii. Most of the remaining patients are referred to the Tripler Army Medical Center in Hawaii (20 cases of a total 103 patients [or 20 percent] in 1995). The cost for each off-island tertiary care referral is capped at $30,000 per year (although the committee heard reports that some patients' bills were much higher). Referrals accounted for 15 percent of the total health care budget in 1995 (PIHOA, 1997).

Health Care Facilities

An 80-bed hospital in Koror opened in December 1992 and is managed by a health services administrator who reports to the minister of health. Although the physical plant appears to be in good condition and relatively well maintained, it does lack some basic equipment. Recently, more than 110 pieces of major medical equipment were purchased under a special grant made available through the U.S. Congress. A room in the hospital was recently designated as the medical library and the telemedicine and telecommunications center. Funding for books and equipment is being requested but is unavailable.

The new hospital provides the anchor for a health care system that also includes four superdispensaries, nine smaller state dispensaries, an ambulatory care center, and a community health center (the old hospital). Each dispensary is staffed with a nurse or a health assistant, whereas the superdispensaries are staffed with a doctor and a nurse, with telephone linkage to the hospital if needed for consultation.

Providing care to the outer islands continues to pose challenges. In addition to the main airport near Koror, two small airports in Peleliu and Anguar are available for patients going to Koror. Some concern was expressed about reaching islands farther out. Although they keep in touch via radio or a ship making an occasional field trip, the islands have no airports and can be reached only by ship.

Health-Related Community Organizations

Palau has several community-based organizations that relate to health care: the American Red Cross, high school group mentors such as Pride, Shalom, Karui el Make er Ngii, the Committee on Population and Children, and the Alcohol and Substance Abuse Prevention Program (ASAP).

HEALTH CARE RESOURCES

Financial

Funding for the health care budget comes from a variety of sources including Compact monies. In 1997 U.S. federal grants and aid from other international donors accounted for $2,350,500—or roughly 20 percent of the total health care budget. Proposed legislation levying taxes on such things as diving, cigarettes, beer, wine, liquor, and canned meats and to provide funding for health care is pending in the national congress. Some private insurance is also available. The Palauan government is, however, considering passage of a Palau National Health Care Plan (NHCP), which would set up a nationwide public health insurance system. Under NHCP, hospital care and preventive care services would be provided to all citizens and would require only a small copayment for outpatient services, emergency room visits, and prescriptions. All resident aliens would be required to enroll in the plan with premiums paid by their employer if they earn less than $10,000 per year and by themselves if they earn more than $10,000 per year. Monthly premiums would range from $25 to $70, depending on the number of dependents.

Workforce[18]

Physicians

Twelve graduates of the PBMOTP are currently working in Palau (one is on educational leave taking postgraduate courses in obstetrics in Fiji). Eighty percent of the physician workforce is native Palauan. Most of the native physicians provide primary care, whereas expatriate contract workers provide specialty care. Palau now has four surgeons: two Palauans who returned after training abroad, one South Korean (supported by the South Korean government), and one contract surgeon from Burma. One Palauan doctor who trained in the United States and returned to Palau has opened a private health clinic and is doing extremely well; patients have been known to come from as far away as Yap for treatment.

Dentists

The supply of dentists and dental assistants is critical. Two of the three dentists are expatriates under contract.

Nurses

Many nurses have left Palau where they make between $6.40 and $8.40 an hour, to work in Saipan and Guam, where the entry pay scale is considerably higher. The Palau Community College helps coordinate continuing education courses for nurses. Since 1992 several classes of Palauan nurses have participated in a distance education course offered from the University of Guam (Fochtman et al., 1997). An on-the-job training program is also offered to nurse's aides and practical nurses at the hospital. They receive a subsidy of $50 per week and are guaranteed a job after completion of the training program.

Other Health Care Personnel

Officials reported a shortage of nurses, pharmacists, medical laboratory technicians, and radiologists. A psychiatrist and a clinical psychologist are also needed.

Traditional Health Practices

The usefulness of herbal medicines and acupuncture is recognized and the Ministry of Health desires their use, but protocols or procedures for

---

[18]Workforce data are from the 1997 PIHOA Data Matrix (PIHOA, 1997).

incorporating them into the local health care system have not worked out yet. Palau participates in the Western Pacific Region of the World Health Organization's current effort to address this issue seriously.

Quality Assurance

The nurses in Palau have established a nurse's association, which helped to create a Nurse Practice Act and a nursing licensing committee, which requires continuing education before recertification. In 1996 doctors began requiring continuing education for themselves, and recently the Belau Medical Society has been restarted as a professional organization for physicians and medical officers. Comprehensive medical licensure legislation for various health professionals is pending in the national congress.

FUTURE HEALTH CARE ISSUES

Palau plans to continue to improve its primary health care system. This will include better equipment at the facilities and more trained staff. Although expatriates will continue to be used for specialty care such as oncology, cardiology, and urology, the government also plans on sponsoring some of the new medical officers so that they can go to Fiji to obtain additional experience in these and other specialties. Careful planning will be needed to ensure the adequate provision of health care services when the capital makes the planned move from Koror to Babeldaob.